CAN OPENERS

by
Nicholas Moran

Published by Echo Point Books & Media
Brattleboro, Vermont
www.EchoPointBooks.com

Can Openers
978-1-63561-859-4 Casebound

Cover design by Kaitlyn Whitaker

10 9 8 7 6 5 4 3

Foreword

Nicholas Moran presents for your edification and occasional amusement (They actually tried that?) the development story of the American tank destroyers (TDs) in World War II, which in my book The Tank Killers, I characterized as one of the most successful failures in military history. The post-war US Army disbanded the tank destroyer force because tanks were now able to kill enemy tanks as effectively, and because the tank destroyer doctrine had rarely worked as anticipated. But, the Army judged, the tank destroyers had been extremely effective in battle as actually employed. This work explains how the equipment that the TD crews used with such skill and daring came to be.

When American troops entered their first offensive operations in 1942, the TDs that went into battle were technical kludges rushed into service. Here you will learn how much worse the situation could have been, and why the vehicles issued to combat troops were the ones that made it through the development wringer. The Army started TD development only in early 1941, a year-and-a-half after Hitler's panzers had overrun Poland. That was a late start in the quest to find weapons and tactics that would stop the panzers, something not even the French and British had managed to do.

Moran's research into the records, primarily those of the US Army Ordnance Department, is stunning in its depth and integration of the story across platforms. You will learn how, from the get-go, the Army tried to create weapons that could shoot and scoot cross-country to implement the doctrine of ambushing enemy armor—and the TD force's preference for speed and maneuverability over armor protection. The range of ideas tested during 1942 alone is amazing. TD development also gave the mechanized Cavalry its workhorse armored car during the war, the M8. Given that other authors have provided detailed works on the service of the M10, M18, and M36 TDs that equipped the TD force from late in the North Africa campaign, this work focuses on the development phase. The near demise in 1943 of the project that produced the M36 with its panzer-smashing 90mm gun, and wisdom of those who kept it alive "just in case," is particularly intriguing.

Granted that today's weapon systems are far more technologically complex, it is still refreshing to see how quickly and relatively efficiently the Army could produce and field world-class tank destroyers at a time when the job simply had to get done.

Harry Yeide
18 October 2017

CONTENTS

Introduction

There have been numerous other books written which more than adequately describe the design and use of the gun motor carriages fielded by the US Army's Tank Destroyer branch. It is not the purpose of this volume to replicate the works of others, though inevitably there will be some overlap. Instead, the focus here is a little more esoteric: It is to bring to light the development process and various prototype vehicles which the US Army specified, designed, built and tested, at least to the stage that any particular carriage made it through that process.

The following descriptions and reports are primarily compiled from the original Ordnance Branch records. Much of what follows is lifted from either the Ordnance Branch's official histories or the project reports. Some poor fellow in the 1940s spent quite some time slaving over a typewriter recording this stuff, it seems a shame to have the end result sitting all but unread in the National Archives. Other significant sources are the proving ground reports, and the reports of the various service boards whose job it was to evaluate the products that Ordnance were sending them before deciding whether or not what the engineers thought was a worthy product in the test facility was actually of any use in the field. As a result, I don't consider this work to be 'written' by me as much as 'compiled.' I have taken a somewhat Jentz/Doyleian approach. If I didn't personally see something in the record, it does not appear in this work.

Photographs which as far as I know have not been published elsewhere have been selected in preference, even if they are perhaps not to the same quality as others which could have been used. I see little purpose in replicating the work of others, and hopefully folks such as modellers will find the new views useful references.

Though the Tank Destroyer Branch is known as being the US Army's signature engine of the destruction of enemy armour, there are some misconceptions commonly held which should be disposed of before delving into the systems themselves.

The major one which should be addressed is that the destruction of enemy armour was by doctrine the purview of solely of the tank destroyer force, and that all purpose-designed anti-tank systems in the US Army were tank destroyers. The latter can be easily disproven by looking at the fact that the 57mm anti-tank gun was in common use. These would be found in the infantry divisions as organic assets. Similarly, US tanks were generally expected to be able to engage any tanks that they came across. Although this detail is only referenced initially in the Armored Force FMs such as 17-10, the later Tank Destroyer FM 18-5 of 1944 also points out that an enemy armoured attack can be defeated by an armored division's organic assets (i.e. tanks).

That the tanks proved unable to do so on occasion (such as demonstrated by the difficulty of 75mm armed M4 Mediums in killing Panthers from the front and Tigers) was no fault of the doctrine, but instead of a failure in intelligence in realizing that the 75mm simply wasn't up to the task that the doctrine required and that even the 76mm needed a hotter round. That, however, would be the subject of a different work. In any case, the motor carriages in this volume were variously tested by Armored Force, Cavalry, Infantry and Tank Destroyer Command for use in their own organisations: The common ground being that the primary design criteria of the vehicles was that they be used to defeat enemy armour.

The role of the tank destroyers was the destruction of massed attacking enemy armour. A review of the field manuals (FMs) issued to the TD units indicates that TDs were supposed to be used en masse, and both during offensive and defensive operations they were to be allocated to engage enemy armoured forces which were either attacking or counter-attacking. It is emphasised in the FM that TDs should only ever ambush enemy vehicles, never hunt or stalk them. As a result, under doctrinal conditions, the TDs never really were expected to take the first hit, so armour wasn't much of a problem. The British would consider such vehicles to be self-propelled anti-tank guns and used them in such a manner, and often it is forgotten that the same FM 18-5 also covered the un-loved tank destroyer battalions equipped with towed guns.

Again, doctrine didn't necessarily match up with reality, though this time it was something of a reverse of the M4 problem: Commanders rarely wanted to let something with tracks, armour and a gun (not to mention four or five soldiers) sit around unused just in case the enemy might happen to launch a massed armour attack. Eventually doctrine evolved so that the updated FM did allow for use of TDs to crack open strongpoints, for example, but this is a bit off-tangent for the purposes of this volume. Suffice to say, however, that a high- velocity direct-fire gun motor carriage suitable for Tank Destroyer use may well have had merit in the other roles of infantry support and vice versa.

It generally happened, thus, that the direct fire gun motor carriage (GMC) was almost exclusively used by the Tank Destroyer units, but it seems to have turned out to be so not so much by fiat as much as by the simple fact that no such GMC was widely accepted for service as being suitable for the needs of the other branches. The US Marines were an obvious exception, using the 75mm GMC M3, though as much as an assault gun, as Japanese armour was not particularly troublesome.

That did not, however, stop various designs being developed for those other branches, and they will appear in the following pages alongside their more famous TD cousins. It was long considered, for example, that the anti-tank companies in the infantry divisions probably needed to be a little more mobile than simply being towed guns as speed of reaction would be important even for localized, small-scale armour problems that didn't merit the deployment of a TD battalion.

It is possible, for ease of categorization, to look at these developments by either gun caliber, or by running gear. To keep it simple, this work will split the carriages up into wheeled, half-tracked, and full- tracked categories first, then progress upwards through the calibres within each type. Though generally the vehicles will be described in chronological sequence, please note the dates, as there was a fair bit of contemporaneous development.

Finally, four acknowledgements. Firstly, to Ken Estes, who helped me find my feet around the National Archives and without whose push in the right direction I would not have been able to get started on this, and to Steve Zaloga for giving me a bit of support, and providing some photographs (particularly of the T101 in testing). Then to Harry Yeide, for the foreword. Finally to Hilary Doyle, who is just generally an all- round good guy.

Wheeled Vehicles

The wheeled tank destroyer program seems to be the least well known of the categories, even to the extent that few will recognise the combat-service M6. Wheeled vehicles showed great potential in the role of a highly responsive and mobile force, though as guns became larger and heavier, they strained the capacity of the chassis, and reduced off-road capability.

37mm

The 37mm anti-tank gun, very loosely derived from the German PaK 36 of the same caliber of which the US purchase two for study, was the primary anti-armour system of the US Army at its entry into WWII. In early 1941 it was proposed that greater tactical utility could be made of a gun which could be fired from its transport vehicle as opposed to the towed variant: The loss of concealment caused by the larger visual signature could be countered by the speed by which the gun could be brought into, and taken out of, action. As a result, two projects were started: The 37mm Gun Motor Carriage T8 and the 37mm Gun Motor Carriage T2. As it turned out, this was going to be the first of many proposals to turn the 37mm into a self-propelled system. It also turned out that this was all something of a lost cause as by mid 1941 the 37mm was as near to obsolete as made no difference anyway.

T2 Series

A nifty little vehicle had been demonstrated to the Army by early 1941 by the American Bantam Car Company: A ¼-ton liaison truck which demonstrated quite superior capabilities offroad. In May 1941, a stripped down example was to be found being tested in Fort Bliss by the 1st Cavalry Division. COL B. Q. Jones, a cavalry officer, designed a pedestal mount for the 37mm Gun M3 which was affixed to the rear frame of the vehicle, allowing a 360 degree arc of fire. More specifically, the gun was fitted to a Bantam chassis from which the body had been removed. A steel plate was mounted across the frame to hold the gasoline tank, which served as a seat for the driver. Another steel plate, underneath the rear of the frame, supported the pedestal mount for the 37mm gun. Initial results were promising, but it turned out that COL Jones was not the only one playing with the idea. Another 37mm on a 4x4 chassis (T8) was soon being tested in Aberdeen, with the significant difference that the 37mm was mounted in a more forward position, facing to the front.

In view of the favourable initial demonstrations of the concept, news of this found its way from the Chief of Cavalry to the Chief of Infantry and finally through the Chief of Ordnance to the Ordnance Committee, the decision-making body which authorized the official creation and nomenclature of Army projects, which authorized the T2 project in OCM (Ordnance Committee Minute) 16802 on 05 June 1941.

Seventeen ¼-ton trucks were to be converted into motor carriages for test by interested arms and Aberdeen Proving Ground. Six were to be built with the gun facing forward in a similar configuration to that which Aberdeen had tested, this was to be 37mm Gun Motor Carriage T2. Eleven were to have the gun mounted on the rear of the truck, in a similar configuration to that which COL Jones had been testing earlier, as T2E1. The mounts were named accordingly, the Aberdeen Mount and the Jones Mount. An eighteenth vehicle was authorized in August 1941.

37mm Gun Motor Carriage T2

This motor carriage retained the main features of the ¼-ton Bantam truck with a minimum of modification. From the standard truck the windshield, the driver's seat, the assistant driver's seat, and the back seat were removed. The gas tank was moved to the extreme left and tank-type removable back seats were provided. A special pedestal mount for the 37mm Gun M3 was welded to two angle irons, which in turn were welded to a steel plate, secured by bolts to the frame. The pedestal mount as designed was fashioned after the Jones mount in that a pintle was used to take part of the M4 Carriage and inserted in the pedestal which was then mounted to the vehicle.

Two views of T2 in its original configuration. In both photographs the gunshield is in the lowered position.

2

Original configuration of T2, with the shield lowered.

Positioned between the driver and the rider, the gun faced forward, firing across the hood of the truck. The original traverse of the M4 carriage, 30° left and 30° right of center, and the original range of elevation, from -10° to 15°, was retained. In order to clear the front of the vehicle, it was necessary to put the antitank gun in such a high position that firing to the flanks would be impossible without turning the vehicle over. In an example of some creative reporting, however, it was considered that the gun had potential 360° traverse in that the driver remained in his seat and could augment the traverse of the carriage by moving the vehicle. A full-width gunshield was mounted, with the upper plate hinged to allow it to fold down to allow the driver to see where he was going.

Six vehicles were assigned to the 93rd Anti-Tank Battalion in Fort George G Meade, Md, for tests, one stayed in Aberdeen. Four of them were brought back to Aberdeen for firing tests, with one of the vehicles suffering a broken frame after the 60th round. When fired, the carriage was reported to pitch very badly.

Right rear view of the original configuration of T2, with the shield lowered.

Front view of the modified T2. The shield remains unchanged, and the steel plate to protect the hood is visible

Since the muzzle was very close to the hood on maximum depression of the gun, a ⅛" steel plate was welded to the hood to strengthen it against cave-ins from muzzle blast after initial attempts at preventing cave-in by welding angles on the bottom of the hood failed. A fire extinguisher was mounted to the right rear rib of the pedestal mount.

After the initial tests, the vehicle was modified somewhat. The original ammunition stowage under the gun was considered unsatisfactory as it could not contain two standard 20-round 37mm ammunition boxes. A new stowage box for the standard boxes was built, opening to the right of the vehicle, with a cut made in the body to allow for it to open completely.

The standard rear seat of the Bantam was removed and modified mounting the seat portion flat on the floor up against the tool box at the rear of the body. The back was mounted in its original position and welded in place. The gunner and loader, increasing the crew to three, would ride on this seat. The standard Bantam seat for the driver and assistant driver as furnished were found to be too wide. Standard tank seats with detachable backs were substituted in their place.

Right side view of the modified T2, now with three crew, as if the vehicle wasn't overloaded enough already

4

The nonstandard gun shield was necessary to protect the driver, to eliminate interference with the steering wheel and so as to not interfere with the driver's movements. No safety belts were provided for the crew of the vehicle. It was felt that the driver could hang on very well. The crew of gunner and loader were crowded in their seat with the guard of the gun and the top bow to hang on to.

Aberdeen also reported that the T2 was a very good riding vehicle, although it felt slightly top-heavy to the driver. This is unsurprising, as the profile view of the vehicle shows. Further yet, the ¼-ton Bantam had a rated maximum gross weight of 2,080 lbs. but the T2 with crew and a full load of ammunition weighed in at some 3,510 lbs: Nearly ¾ of a ton more than the ¼ tons of cargo that the truck had been designed to carry.

Top: T2 Modified. Left: The pedestal during the modification process. Above: One of the T2s used in maneuvers in Maryland.

Details of plate and mount on 37-mm Gun Motor Carriage T2E1, first pilot. This mount had a theoretical traverse of 360°.

Above: T2E1 in its first iteration after the pilot. Note that the steering wheel is angled forward compared to the original vehicle below.

The Jones mount was shipped from Texas to Aberdeen and used in building the pilot T2E1. The Jones pedestal on the T2E1 was supported by a new plate, made for the back of the frame, and the gun was mounted approximately 4 inches higher than on the original vehicle. Originally COL Jones had mounted his pedestal on the underside of the frame by removing an "X" cross-member. This mount had a theoretical traverse of 360°, with an arrangement of four bolt heads and pins at the top of the pedestal, for locking and unlocking the pintle.

Left and below: Two views of the vehicle undergoing conversion to T2E1, showing well the Jones Mount

With these bolts locked, the gun could be traversed only through the conventional 60°; by loosening them, the gun could be turned through any angle. This was more theoretical than actual, however, as it is immediately obvious from a cursory glance at the photographs that the forward arc could only be entered with the gun at high angles of elevation. On the plus side, the center of gravity was lowered considerably compared to that of the T2 which made the whole system more stable than T2 when firing directly to the rear, as well as presumably making life much easier for the driver, particularly on side slopes.

Left side view of the T2E1 undergoing conversion

Another advantage of the T2E1 over the T2 was that of weight: The vehicle tipped the scales at a 'mere' 3,250lbs, with its crew and 40 rounds, but the majority of this is accounted for by the fact that there was only a crew of two, the loader performing double duty as a driver. There was less requirement for a driver to be ready to act at all times on T2E1, as the gun could be fired to the flank, although the recoil did cause the tyres on that side to lift off the ground.

Two views of T2E1's 37mm gun being serviced.

After the first pilot vehicle, subsequent models had the driver's position moved forward, with the steering column tilted 15 degrees forward to compensate. Records are unclear if the pedals were moved forward as well. The driver was now equipped with a collapsible tank seat, whereas the gunner's seat faced the rear, towards the gun and thus on the correct side of the cannon to control the gun when firing to the rear.

T2E1 with the revised rider's position.

The rider was considerably cramped in this riding position. This was caused by the fact that it was he had to have his feet underneath and in between the gun control mechanism, or else touching the fender which was vibrating as the wheel moved up and down going over bumps. This would not only result in an uncomfortable position for the rider but, if the rider's leg were to touch the fender the result would probably be a seriously bruised leg.

As an alternative position for the gun and rider, the gun was turned around. With the gun carried in this position the rider had no interferences at all from the gun mechanism and could sit comfortably resting against the shield, have his left arm resting on the gun and his leg could stretch out or prop up against the ammunition box. The rider's seat was also turned around, it was necessary to make a minor cut which was also to allow complete lowering of the breech and thus complete elevation of the gun.

Left: Servicing the 37mm to the rear appears to have been an uncomfortable affair. Below: T2E1 with seats stripped and ready for firing.

On the back of both the driver's seat and the rider's seat was placed a bracket for the safety belt. The bracket was welded rigidly to the rider's seat and was welded on a hinge which was mounted to the driver's seat. It was necessary to mount the driver's safety belt on a hinge so that it could be folded out of the way and would not interfere with the breech of the gun when it was traversed and elevated. In order to set the carriage for firing, the seat backs and seat cushions had to be removed.

Unsprung mud guards were mounted on the axle over the rear tires, requiring less clearance than mud guards mounted on the body, with springs. A spare tire was mounted on the hood. A new type pedestal was designed, approximating that of the T2 mount, and equipped with a special locking device similar to that of the Jones pedestal to give free swing to the gun for 360° traverse. A spring-loaded pin, located just below the gun carriage and at the top of the pedestal, locked and unlocked the pintle.

Left: It seems unlikely that if fired from this position, the gun would hit anything other than the spare tyre. Hopefully only an optical illusion.

A further development of the concept was recorded as 37mm Gun Motor Carriage T2E1 (Modified). This was an attempt to simplify the T2E1. It used the pedestal designed by Aberdeen Proving Ground and had the characteristic 60° traverse of the M4 carriage. The main advantages of this model were: it was more easily manufactured; it retained the original body; it was more stable; it could retain the windshield due to no requirement to move the steering wheel into where the windshield would ordinarily go; and the pedestal could be mounted in the rear, as was done in the original Jones mount and the T2E1. The gross weight was 3,670 pounds, making this vehicle, like the earlier modifications, an overload for the Bantam chassis

Left: T2E1 from the rear. Below: T2E1 (Modified)

Above: Two views of T2E1(Modified).

Above: The new mount which raised the gun
Below: Right side view of T2E1(Modified)

An Armored Force Board report dated 29OCT refers to "T2E2 (Jones Mount)". Exactly how this links to the T2E1 is not made clear. It may be the T2E1 (Mod). Extracts from this report are as follows:

Discussion:
a. A total of 52 rounds 37mm AP ammunition was fired from the mount at both fixed and moving targets. The crew consisted of two experienced sergeants of the Anti Tank Company of the 46th Infantry. Two hits were made out of 10 rounds fired at moving targets at the maximum range of approximately 1,500 yards and a minimum of 425 yards. Several hits were made on fixed targets at 500 yards range.

b. Full motion pictures taken during firing show that each time the gun is fired considerable rocking and bounding occur . Each time the gun is fired broadside, which is the easiest position to serve and fire the gun, the entire mount displaces from four to five inches. The rocking and jumping of mount prevents the gunner from remaining close to the sight in attempting to track the target. The sight is designed for an eye position 6" from the sight; with this mount the gunner must keep his eye at least 18" away.

c . The time required to move from a traveling position on a road to a firing position 150 yards off the road, was 60 to 70 seconds. This time included the laying of the piece and firing the first round. To go into a firing position in the direction of travel required 15 to 25 seconds. The time required to go into a new position with the gun in firing, position and fire a round varied from 25 to 55 seconds depending on the distance traveled (20 to 125 yards).

Left: Assistant Secretary of War J.J. McCloy and Major C.C. Drury, Canadian Army, inspect a T2E1 in Ft Meade. Above right, T2E1

The Armored Force Board reported to the Chief of Ordnance that the Jones mount overloaded the ¼-ton 4x4 truck by about 500lbs, that the mount was not sufficiently stable during firing to allow the gunner to track the target accurately with the aid of the sight, and that the provisions for traversing the piece were not satisfactory (i.e. too limited).

The Armored Force Board recommended that the 37mm gun have a minimum crew, including driver, of three men and preferably four, which wasn't possible on the vehicle, and stated that a similar mount on a heavier vehicle of the same low silhouette would undoubtedly be more stable and more satisfactory.

Accordingly, the project was terminated in OCM 17643 of 14 January 1942, with the recommendation that future attempts to make a 37mm GMC be restricted to vehicles of longer wheelbase and which were designed to carry the weight of load to begin with.

The rider of this T2E1 is using the gun as a point of support as the vehicle traverses uneven terrain

37mm Gun Motor Carriage T8

First Ford pilot, T8

The effectual contemporary to T2, T8 was the first 37mm GMC built in Aberdeen. In this case, the weapon was mounted onto a specially-designed vehicle known as the "Swamp Buggy."

Above: The original Aberdeen pilot.

Manufactured by the Ford Motor Co., at the direction of the Commanding Officer, Holabird Quartermaster Depot, the chassis was composed of standard Ford 1½-ton components together with a special frame and body. The engine, a Ford 6-cylinder inline L-head, was housed in a mounting forward of the rear axle, to the right of the driver's seat.

Left: The mounting for the 37mm as found on the first Ford vehicle. Right: Front view of the first Ford pilot in the field

Two additional seats were provided for the gunner and loader, one at the right, in front of the engine housing, facing forward, and the other behind the driver's seat, facing to the rear. Purchase of two such chassis for use in the development of a 37mm anti-tank self-propelled mount was authorized by the Secretary of War in June 1941 and was designated T8 by OCM 16835. Procurement of 15 additional chassis was authorized by the Ordnance Committee on 19 July.

Given that the engine was located midships, this permitted mounting the gun at the front. Mounting the 37mm gun on the first pilot at Aberdeen Proving Ground was a rush job (the reason for the haste goes unrecorded) and no detailed plans were made before the gun was mounted. The 37mm Gun Carriage M4 was separated at the trail axle and put on by means of readily available angle irons, which were bent to shape and welded in place on the vehicle. Later a shield, which would fold up or down, and an ammunition box, were added. With just the gun and shield, the vehicle weighed 4,520 pounds; when one added 100 rounds of ammunition, and three-man crew the gross weight was increased to approximately 5,430 pounds. Larger ammunition boxes were later added.

Above: New ammunition boxes for the #1 pilot
Below: T8 Ford Pilot #1 in the field.

The original traverse and elevation of the 37mm Gun Carriage M4 were retained, affording 30° traverse to the right and 30° to the left, and elevation from -10° to 15°. Tests showed the vehicle to be very steady when the gun was fired: It settled on its tires, but did not pitch appreciably. The driver was able to stay in his seat at all times to help turn the vehicle toward any new targets which might appear. He was also able to drive away as soon as the firing mission was completed. The vehicle was considered to have exceptionally good cross-country ability, doubtlessly helped by the large wheels and short wheelbase. However, problems with the cooling and steering systems were identified, and the first pilot was returned to Ford in September 1941 for these to be rectified.

Above: T8 Ford Pilot #1 rear view

The first production model made by the Ford Motor Company was patterned after the vehicle constructed at the Proving Ground. One significant difference was the replacement of the expedient gun mount with a single casting. The rear seat was moved further to the rear, the wheelbase was extended about 4", and the tread 3".

Above: T8 Pilot #2. Note the radiator shroud and seat added behind the front wheel.

The second pilot incorporated a few changes. A shroud was placed over to radiator/air intake to increase cooling effect, and a newly-designed gunshield was added. The steering was improved, a new seat added behind the fender, and the rear seat raised. Although the gun mount now had brackets for a travel lock, they were poorly placed and had to be modified by Aberdeen. Further, the driver's seat was replaced by a collapsible tank seat, and raised to 13.5" above the floorplate.

Left: Two views of the gunshield designed for the vehicle on Pilot #2
Below: The 100-round ammunition box and battery compartment on Pilot #2

In the meantime, the next fifteen vehicles were completed and at Aberdeen by November 1941. Further changes were made, most notably the new gun mounts to allow 360° traverse. This afforded elevation from from -10° to 15°, except for over a 60° section to the rear.

Above: T8 in its final configuration with three seats, 360 degree mount, and side-opening ammunition box

Left: The new 360 degree mount. The handwriting was found on the original in the Tank Destroyer Board archives. Bottom left: Front view of the final T8, with a simplified gunshield. Bottom right: Upper rear view, same vehicle.

Unfortunately, the new mount was susceptible to clogging and binding in its exposed position. The new pedestal was adequately strong, but, together with its mounting plate, was heavy.

By this point, the driver's seat was made adjustable to allow maximum visibility, and the side seat support removed. The ammunition case was also modified, so that it opened to the right. Six of the vehicles were shipped to the 93rd Anti-Tank Battalion, and two found their way north for testing by the Canadian Army. Given they were sent without guns nor cradles, it seems reasonable to presume the Canadians intended testing with their 2-pounder anti-tank gun.

The remainder were tested at Aberdeen and undertook comparative trials with the 37mm T21 described below and T2/T2E1 above. Cross country mobility and firing remained good, though heavier shock absorbers were recommended for the front suspension. The traveling lock was not used as it was a bit inconvenient in combat, but was considered desirable for protection of the traversing and elevating mechanisms if the vehicle should be used for extended cross-country operation.

Left: New, side-opening ammunition compartment
Below: T8, final variant, with shield up

In comparison with T21, it was found that: "In general it was found that the cross-country ability of T-8 and T-21 is about equal, the T-8 able to travel over rough terrain a little better but the T-21 able to ford to a slightly greater depth. Visibility is slightly less on the T-8 than the T-21 and time necessary to get into action and out again is approximately the same on both." As an aside, this may be an opportune time to note that though the US nomenclature system did not use dashes, it has proven not uncommon to find them in official documentation regardless. It evidently was not a major point of concern.

In independent testing, the following observations were noted, amongst others:

With the present arrangement of control pedals, a driver with comparatively short legs is unable to take full advantage of the seat provided. This is a decided disadvantage and must be corrected.

Fuel consumption, as based on the overland strategic run has been determined at approximately 10.7 miles per gallon. The 50 miles cross country operation immediately following the strategic run resulted in a gasoline mileage of approximately 8.5 miles per gallon.

Sand performance of this vehicle is considered excellent in first gear. The 9.00x20 tires add considerably to the flotational qualities of the vehicle. The combination of comparatively light vehicle weight and large section tires makes this vehicle exceptionally well suited to both sand and mud operation.

Safety belts have not been provided for the crew, and therefore, in cross country operation they are subjected to undue jostling. Due to the absence of a windshield, road speeds in excess of 40 miles per hour for any length of time requires special clothing for the crew if the outdoor temperature falls below 35°. On the continuous operation tests, several stops were required to enable the crew to warm themselves.

Final version T8, shield down

The Armored Force Board, after its own testing in January 1942 concluded that with certain proposed changes the T8 was a "fairly satisfactory self-propelled mount for the 37-mm antitank gun," but that in view of the standardization of the 37mm Gun Motor Carriage T21, the T8 should not be considered further. Ordnance Committee action in OCM 18127 on 23 April 1942 accordingly directed that the project be closed.

By this time, four of the vehicles had been sent to Canada and two to England (If the additional four vehicles were armed or not is not known to the author at this time); the nine still available were returned to Aberdeen Proving Ground for removal of the guns and use of the chassis as proof facilities.

Two views of the final version of 37mm Gun Motor Carriage T8

37mm Gun Motor Carriage T21 (M6)

37mm GMC T21, fully stowed

On 13 June 1941 Robert Biggers, president of Fargo Motor Corporation, a subsidiary of the Chrysler Corporation, submitted to Colonel Christmas of the Ordnance Department a "very crude" model of a proposed 37mm gun motor carriage utilizing the ½-ton 4x4 Fargo truck (later classified as ¾-ton 4x4 Dodge truck), which was being supplied in quantity to the Quartermaster Corps.

The advantages pointed out were:
(1) the gun had a traverse of at least 270° without interference with the driver;
(2) there was adequate working space for the gun crew and protective armour for the gunner without obstructing the view of the driver; and
(3) the gun was carried inboard, where it would be protected from trees, brush, and other obstacles and could be manned for instant action at all times.

The model sent to Ordnance

The model had the driver's seat moved to the left as compared to the standard vehicle, but Mr Biggers did mention it would only be done as a last resort. A particular advantage cited was the fact that the basic vehicle was already in wide use by the Army, simplifying service and supply.

A full-size mockup of the proposed vehicle was delivered to Aberdeen Proving Ground in August 1941, and the 37mm Gun M3 and pedestal were mounted on it.

The mock-up with the gun mounted in the two positions considered

For comparison, the vehicle was tested with the gun mounted first facing front and then rear. With the front mount, the driver faced the scene of action and could maneuver the truck with the gun in firing position. A further advantage was that front mounting distributed the weight of the gun and mount more evenly over the vehicle. On the other hand, front mounting had three disadvantages: (1) it necessitated turning the vehicle to leave the scene of action; (2) the car and driver were affected by the muzzle blast; and (3) it was difficult to construct a suitable shield that would not interfere with the driver and the steering wheel. With this mount, the gun had to be placed high enough to clear the hood and provide 10° depression, necessitating an increase in trunnion height of 10 inches over the rear mount. The windshield had to be discarded.

With the rear mount, the vehicle could leave the scene of action without losing time to turn around. In this position, the gun had sufficient clearance above the sponsons to permit 360° traverse, permitting emergency firing to the front and to the frontal flanks, and the full elevation of 15° and full depression of 10° were available through an arc of approximately 300°. With the gun shield in the rear, the driver had better vision and the standard windshield could be retained. The standard gun shield of the 37mm Gun Carriage M4 could be used. With the rear mounting, however, the truck would have to be turned around when entering the scene of battle and it would be more difficult to maneuver the vehicle into firing position while backing up. The weight of the gun and mount would be entirely on the rear axle. Aberdeen Proving Ground recommended that the gun be mounted for firing normally to the rear. In accordance with this, Ordnance Committee action in September 1941 (OCM 17273) designated the vehicle 37mm Gun Motor Carriage T21 and directed that the gun be mounted to the rear.

The first pedestal used was the same as that employed on the 37mm Gun Motor Carriage T2, except that two ribs were added for greater strength. It afforded 360° traverse by the use of a large gear with an external pinion operating from the standard traversing mechanism of the 37mm Gun Carriage M4. The cannon mounted was the M3.

Initial tests starting July 11 were focused on mobility, with the vehicle performing well other than its inability to climb a 12" step. "During the short time the vehicle was tested, riding qualities were considered excellent", as was its performance in sand in low gear. There was some consideration given to replacing the Chrysler seats with tank seats to reduce interference with the gun, but it was concluded that the better solution was simply to cut down the tops of the seats.

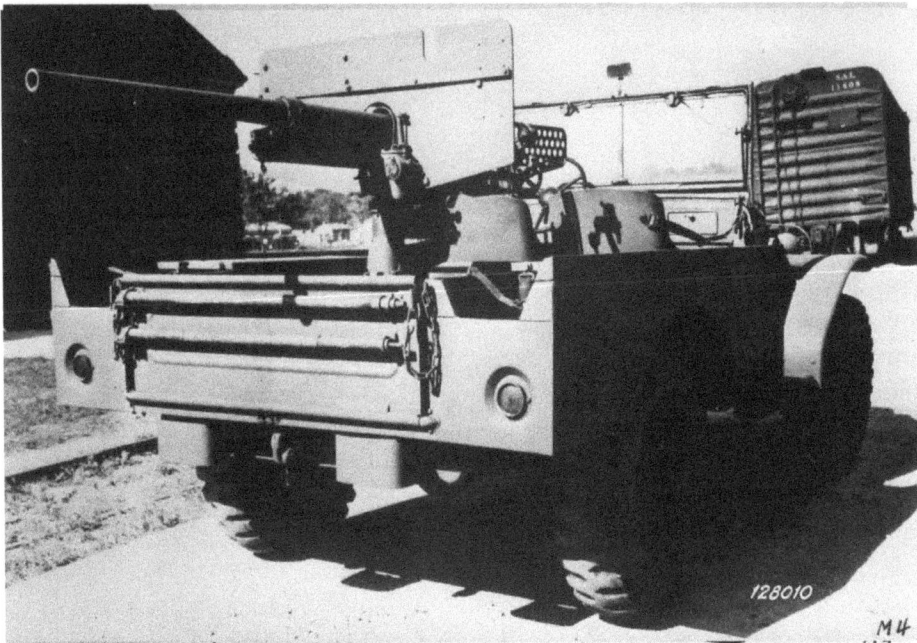

An initial firing program of 200 rounds was initiated, with the vehicle's integrity suffering only the loosening of a few bolts on the body. However, firing to the front proved a little more destructive:

a: With the gun pointing directly to the front and at zero elevation the muzzle blast of the gun knocked two instruments loose on the dash board and shattered the glass on all five instruments. The dashboard is approximately 18 inches in front of the muzzle. No damage occurred to the windshield though only one round was fired at zero elevation.

The original pilot model with the shield from the towed carriage

No damage resulted to the instruments when the gun was fired at maximum elevation in this position.

b. When firing directly to the front at maximum and zero elevation the radiator cap blows off. This was due to the radiator cap having a loose fit and not seating properly.

Other failings noted in the first report dated 8th November 1941 were that the standard shield from the M4 carriage was too thin and small to adequately protect the gun crew from small arms fire, no protection was provided from aircraft fire at all, and there was no provision for automatic weapons for the crew in case the vehicle was disabled. Overall, however, initial results were promising, and the report recommended the "a few of these vehicles be procured for use as 37mm gun motor carriage and subjected to a thorough field test in the using services".

The pilot was sent to Fort Meade, Md., in September for test and study in conjunction with tests of other 37mm gun motor carriages by the 93rd Antitank Battalion. Further testing took place at Aberdeen through October, November, and December 1941, to determine the deficiencies existing in the pilot vehicle.

Left: The new travel lock. Right: Pilot model pedestal

In November Aberdeen designed a new shield to provide frontal, semi-overhead, and flank protection against cal. .30 ball ammunition at all ranges. At the same time, a new travel lock was designed, but the pedestal and shield mount were considered inadequate.

The replacement pedestal designed by Duplex Press, now combining the travel lock with the enclosed gearing. Note that the gunshield no longer is flush to the sponsons

Three new pedestals, varying principally in the sturdiness of their construction, were manufactured by the Duplex Printing Press Co., Grand Rapids, Mich. The heaviest was used. The Duplex Co. also provided a fully enclosed gear train for the traverse and redesigned the shield support.

Aberdeen's later report on the T21 recommended its adoption as Standard, provided the engine cooling could still be improved. The Proving Ground did not consider this vehicle as good as the 37mm Gun Motor Carriage T8, but recommended that it be used as an expedient because of the availability

of the Fargo chassis, which was in production commercially. The Aberdeen report recommended the following improvements:

1) Because the gunner's shield impaired rear vision, adequate mirrors were needed.
2) The scoop effect of the shield directed a draft on the occupants of the two front seats, and to overcome this a canvas tarpaulin should be lashed from the breech of the gun to the shield.
3) The ammunition wells were to be deepened slightly to facilitate removal of ammunition boxes, and the shield altered slightly to accomplish this.
4) If possible, a positive locking device was to be incorporated to engage the front wheels when operating in low gear, thereby eliminating excessive strain on the rear axle.

A comparative test of the 37mm Gun M3 and the 37mm Gun M6 was made to determine the value of the semi-automatic feature of the M6 gun. This test revealed that the loading time was not a serious factor in the rates of fire and that a complete gun crew could fire as many aimed shots from an M3 gun as from an M6. With a reduced gun crew, the M6 could be operated more rapidly than the M3, but this advantage was offset by the sensitivity of the M6 gun to powder pressure variation and difference in recoil. The Proving Ground recommended that

The new gunshield had cutouts to provide suitable clearances at varying degrees of traverse. Aberdeen reversed this on the first production vehicle. Here you can see the removed parts have been welded back on.

until such time as the M6-type semi-automatic gun was developed to operate satisfactorily with all types of ammunition and at any temperature, the M3 hand-loading gun be considered superior. For future gun motor carriages of this general type, Aberdeen recommended consideration of a pedestal using an x-type frame integral with the vehicle frame. It further recommended provision of a gunner's seat that would traverse with the gun, open sights for use when the optical sight became inoperative because of dust or dim light, and a blast deflector to prevent the raising of dust in front of the muzzle.

A preliminary report from the Assistant Chief of Staff, G-3 (Operations), to the Deputy Chief of Staff comparing the 37mm Gun Motor Carriages T21 and T2 gave the preference to the T21. It was easier to drive and provided better visibility for the driver and more comfort for the crew. The T21 also had increased ground clearance at center, its performance in water was better, and it was more adaptable for cargo and personnel carrying purposes. Another desirable feature of the T21 was that ammunition could be carried in standard 20-round packing boxes. The T21 was recommended for standardization by Ordnance Committee action in December 1941. Gun motor carriages were urgently needed in quantity at that time, and since the weaknesses of the T21 in engine cooling and axle torque capacity were outweighed by its interchangeability of parts with the standard Quartermaster ¾-ton truck already under procurement, standardization was approved and the vehicle was cleared for procurement on 26 December 1941.

By February 1942, the vehicle, now designated 37mm Gun Motor Carriage M4 had found its way to Fort Knox. By this stage it had covered over 10,000 miles, and fired over 500 rounds. "The vehicle and gun showed no ill effects as a result of this usage." After Armored Force had finished with it, it was then handed over to 701st TD Battalion for further training and testing.

The final stowage plan for the gun box, first aid kit and two rifles

The designation was changed in February 1942 to 37mm Gun Motor Carriage M6, to avoid confusion with the Medium Tank M4. At the same time the designation of the gun mount was changed to 37mm Gun Mount M25. The first production vehicle, without gun, was sent to Aberdeen Proving Ground in March 1942 in order to work out stowage. When standard ammunition boxes with metal interliners were placed in the wells provided in the vehicle body, serious interference occurred between the tops of the boxes in the two rear compartments and the end of the shoulder guard at the rear of the gun when the gun was placed in the position of maximum elevation.

This interference was corrected by the use of special metal ammunition boxes several inches shorter than the standard wooden box. The shells were packed in fiber containers in these special boxes. There were no other stowage problems.

Top: Side view T21
Above: As part of the testing process of the M24 pintle, it was fitted to T21. Note the gunshield has not had the corners cut.

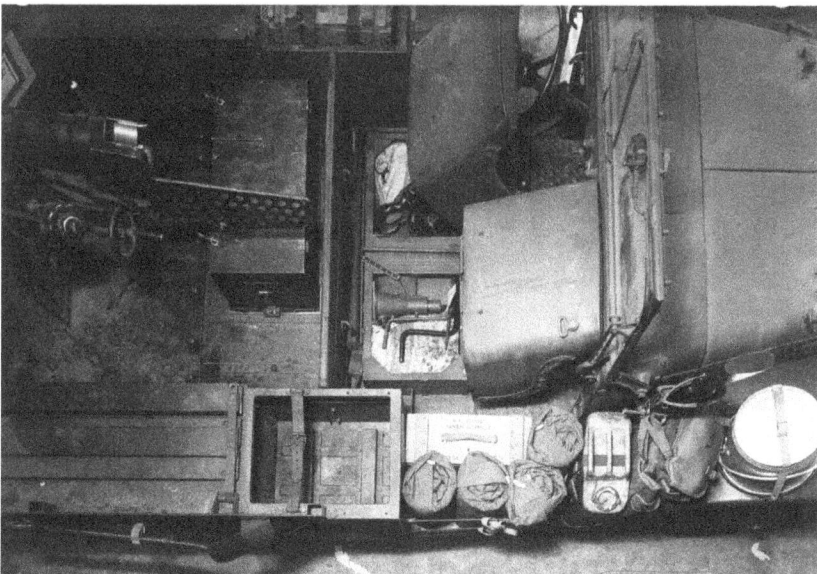

The first production vehicle was tested at Aberdeen April through September of 1942 after an overland delivery, with 773 miles on the clock, 7.7 miles to the gallon. Before the tests started, it was then driven to Fort Bragg for testing by the Field Artillery board, therefore before any serious testing was done at Aberdeen it had driven 1206 miles.

The vehicle was reported to have better engine cooling than the pilot model, but its slope-climbing ability was reported not as good. Indeed, a number of differences in power were noted in the report, and given that the engines had the same displacement and compression ratio, the report theorised that a different octane rating of fuel was used when the T21's characteristics sheets were drawn up compared to the 70-72-octane fuel used for M6. However, the slope climbing difference was attributed to a gear ratio change. The rear differential of the production model was considered too weak to transmit the full torque of the engine when operating in first gear. Aberdeen had, as above, recommended that an interlocking mechanism be incorporated in the transmission case to engage the front wheels when operating in low gear, but this had not been done on the first production vehicle. Such a mechanism, it was thought, would eliminate excessive strain on the rear axle, and reduce the possibility of breakage when transmitting full torque in low gear. During the testing process, the production vehicle suffered only one mechanical failure: During the drawbar pull test, the rear differential spider gears failed.

Two modifications were made to the vehicle by Aberdeen: The seat backs for the gun crews were shortened in order to clear the gun shield. The other was restoring the shield corners.

Left: T21 fully stowed, with tools and chains under the seats

Some other extracts from this report:

The vision from this vehicle is considered excellent to the front and sides. The operator can see the terrain approximately 15' ahead of the vehicle. Vision to the rear is impaired by the gun shield, however, a rear view mirror is installed which is satisfactory for traffic operation.

The windshield wipers have been found to be inadequate for inclement weather operation. They clean only the upper one third of the windshield and the total angle wiped is approximately 80°, which is entirely too limited for safe operation.

This vehicle has a comparatively low broken line silhouette and is easily concealed by the natural ground features. The most conspicuous part of the vehicle is the large gun shield which extends considerably above the highest part of the vehicle. The vehicle operates comparatively quietly over all types of terrain, and raises only the normal amount of dust in dusty terrain operations.

Above: 37mm GMC M6

The riding qualities of this vehicle are considered rough. This is due to the stiff suspension springs and rubber blocks mounted on the chassis frame above the front springs. These blocks limit the depression of the front springs to 1½ inches. Thus, during rough terrain operation, the crew is subjected to considerable jostling and must traverse such terrain at slow speeds.

The high ground clearance of 11" is of primary importance when traversing deep mud. However, when the vehicle mired due to the loss of traction, the winch was found to be satisfactory for self-removal of the vehicle from such conditions.

From the safety standpoint, safety belts and hand grips are provided to prevent the crew from undue jostling. The comfort of the crew is satisfactory except as follows:

a. The driver's seat on this vehicle is not readily accessible for drivers that weigh in excess of 170# and are of normal height. This is due to the low steering wheel and the sides of the seat. The seat is not adjustable and for a driver to get into the seat without squeezing himself, he has to more or less seat himself on the back of the seat and then slide into sitting position from the rear of the steering wheel. To enter the seat from the side, the driver has to get into a semi-squatting position and then squeeze himself between the seat side and the steering wheel. Then for a driver of normal height to put his feet on the clutch or brake pedal, he has to raise his knees up and to the side of the steering wheel, which is an awkward position.

b. The accelerator pedal is located such that it is difficult for the driver to keep his foot on the pedal. The pedal does not incorporate a heel well or plate and is mounted above a convex surface of the floor. Thus, when operating over rough terrain, the driver must keep a strain on his foot so it will not slide off the pedal. However, it is believed that if a heel well or plate would be installed, this deficiency would be eliminated.

The air-scooping effect of the gun shield, which directed a draft onto the front seats, still remained as an objectionable feature. The previous recommendation for the use of a canvas tarpaulin, from the gun breech to the shield, had not been acted upon.

Nevertheless, the Proving Ground recommended that the first production vehicle be considered a satisfactory self-propelled mount, and suggested the following modifications: making the driver's seat adjustable horizontally; installing a windshield wiper with a 10-inch blade to operate through an arc of at least 120°, designing a suitable top to cover the top and rear of the driver's seat; installing a heel plate or well below the accelerator pedal; and bracing the front bumpers adequately on the bottom sides.

Left: M6 with gun canvas. Right: M6 with gun canvas and canopy

Canvas patterns were designed for a gun cover to provide protection for stowage on the rear of the shield and to be an over-all cover for the gun. At the same time, a canopy was designed, stretching from the top of the gun shield to the top of the windshield, for protection of crew members. A recoil guard, to keep the gunner's body out of the path of recoiling parts, was constructed at Aberdeen and proposed for inclusion as an added safety feature.

Aberdeen proposed to apply a blast deflector to the 37mm Gun M3, such a device was designed and tested and reported on favorably. Called a gas deflector, it was actually a form of muzzle brake and increased the stability of the carriage, particularly when firing at right angles to the vehicle. Not that M6 was particularly unstable to begin with.

Aberdeen claimed that in addition to reducing the flash and the amount of dust raised when the gun was fired, the gas deflector protected the windshield, rear view mirror, dash instruments, and steering wheel from the effects of the concussion, making it possible to fire forward over the windshield at elevations as low as 2°. Without the gas deflector, the blast from forward firing at elevations of less than 15° resulted in breakage of the windshield and damage to the panel instruments.

M6 front view with canvas and canopy

In October 1942 the Ordnance Committee designated the gun with gas deflector as 37mm Gun M3A1, and directed that all M3 guns allocated for use on the 37mm Gun Motor Carriage M6 be converted to the M3A1 status. The 37mm Gun Mount M25, when provided with modified firing controls and shoulder guard assembly, was designated as the M26. In addition, it mandated that all future M3s produced be M3A1s.

This caused some concern over at the Tank Destroyer Board, given that they had just recommended that the Light Armored Car M8 be equipped with the hand-operated M3 as opposed to the semi-automatic M6. This meant that whether Tank Destroyer Board wanted them or not, the M8s would be equipped with gas deflectors. As a result, under the instructions of a Colonel Mongomery, the Tank Destroyers conducted comparative trials between the M3 and M3A1 to find out for themselves. 37mm GMC M6s were used for this purpose on 17 November 1942. The ammunition used was APC M61.

Above and below: Comparative blast testing on loose caliche with and without the deflector

1. There is very little difference in blast effect of the two guns over three types of soil: An old plowed field, over a bed of loose caliche, and on a moderately dusty road.
2. If anything, the deflector seemed to increase the total amount of dust raised, although this was kept perhaps on a slightly lower plane in the firing with the gas deflector than without the gas deflector.

3. At no time during any of the firing did the blast obscure the moving target (range about 500 yards) from the gun commander's or gunner's view, either with or without the gas deflector.

Recommendation: That no action be taken to secure the gas deflector on the Light Armored Car M8 (We will probably get it in any case, since it has been standardised on M3 37mm gun)

In addition, the deflector met with an objection from the Army Ground Forces because tests at Aberdeen Proving Ground of the gas deflector with 37mm Tank Guns M5 and M6 showed that the deflector was readily damaged by the use of canister ammunition. A letter from Headquarters, Army Ground Forces, to the Commanding General, Services of Supply, directed that no 37mm gun in the Army Ground Forces be equipped with the gas deflector, but requested the continuation of efforts to develop a device which would reduce dust, protect the personnel and instruments from muzzle blast, and at the same time permit the use of canister ammunition.

Because of its lack of armour protection and other limitations, 37mm Gun Motor Carriage M6 was considered as an expedient only. As the result of the development of Light Armored Car M8, which was considered a more satisfactory combat weapon, the Ordnance Committee, in September 1943, approved the reclassification of 37mm Gun Motor Carriage M6 as Limited Standard. The committee directed that all such vehicles in stock at depots in excess of 100 complete gun motor carriages be converted into ¾-ton weapon carriers, with winch, and that production of ¾-ton weapon carriers be reduced by an equivalent number.

The 37mm guns removed from the gun motor carriages were turned over to the Maintenance Branch, Field Service, for disposition.

Right: Comparative blast testing, ploughed field

Below: Side view of the M24 pedestal being tested on the M6

37mm Gun Motor Carriages T13, T14

Side view, 37mm Gun Motor Carriage T14, first pilot

The idea behind mounting a 37mm AT gun on a 6x6 chassis had been originally brought up in passing in a meeting of 20th May 1941 between a number of representatives of private industry including Mr D. J. Roos of Willys Overland and a Major McAuliffe of G-4. After this initial proposal, tentative drawings of such a Willys 6x6 with the 37mm had found its way to the Chief of Infantry by mid-June. The Chief of Infantry then forwarded them on to G-4 with his endorsement and a suggestion that some pilot vehicles be trialled. This was officially recommended by the Ordnance Committee on 02 July with the instructions that two vehicles be procured: One with a forward gun mount, and one with one towards the rear. This was approved on 03 July.

The first vehicle, designated 37mm Gun Motor Carriage T13, was to have the gun mounted toward the front; 37-mm Gun Motor Carriage T14 was to mount the gun toward the rear. The proposed T13 was never built, because the using arms indicated a preference for a vehicle with the gun facing toward the rear, and Ordnance Committee action in February 1942 authorized building a second T14 in its place.

The first pilot, 37mm Gun Motor Carriage T14, was built by the Willys Overland Co., and delivered to Aberdeen Proving Ground in January 1942.

The chassis differed from that of the ¼-ton 4 x 4 truck by the addition of a third axle, forming a two-axle bogie on the rear.

Rear view, second pilot, 37mm Gun Motor Carriage TI4

A longer, stronger frame was used, and the over-all gear ratio was lowered from 4.88:1 to 5.23:1. Power was furnished by a Willys 4-cylinder, L-head gasoline engine.

Folding seats were provided for the driver and assistant driver, with cushions to the rear of the driving compartment, one on each fender, for the gunner and loader.

Three-quarter left view, 37mm Gun Motor Carriage T14, first pilot.
Note the cushion folded over the second axle

The 37mm Gun M3, mounted in the rear over the center of the bogie, could be elevated from -10° to +15°, and had a 360° traverse, although the normal position of firing was to the rear. It was equipped with a travel lock to keep the gun steady when the vehicle was in motion. The body was designed with a compartment in the rear of the gun mount to accommodate two frames of 37mm ammunition, and a compartment in the gun mount base, opening to the front, to accommodate 26 rounds in racks. The fuel tank was mounted in the extreme rear center of the chassis, forming one side of the rear ammunition compartment. The vehicle was unarmoured except for a gun shield of ¼-inch armour plate.

Side view, 37mm Gun Motor Carriage T14, second pilot, with windshield, top and flaps

After being given a limited firing and automotive test at Aberdeen Proving Ground, the first pilot was delivered to Fort Benning, Ga., in February for a test by the Infantry Board.

The report of the Infantry Board recommended that, with modifications, the T14 replace the 37mm Gun Motor Carriage M6 as a self-propelled mount for antitank artillery. Modifications included: reduction of the height of the gun shield to lower the silhouette; a change in the mounting of the gun to allow for a normal traverse of 360°; improvement in the carburetion, which was insufficient on slopes greater than 40%; installation of run-flat tyres with mud and snow tread (mainly because they couldn't find a place to put a spare, other than on the hood); an increase of the fuel tank capacity from 15 to 20 gallons; installation of a "quick release" lever for the fan belt, and of foot wells for the driver; elimination of jacks, curtains, and top; and reduction of the height of the back of the driver's seat so that it could be folded forward without striking the wheel.

Above: Above right view showing right seat and steering wheel in low position

Right: Frontal view of T14, second pilot with the windshield, top and flaps fitted, 20 March 1942, Aberdeen Proving Grounds

The only adverse comments made by the Infantry Board involved the more or less theoretical factors of unit ground pressure and power-to-weight ratio, high silhouette of the gun shield, and the controversial question of whether it was vital to fire the gun directly over the front end of the vehicle. These factors were considered of slight importance in view of the maneuverability of the vehicle. It was suggested that a blast deflector applied to the gun muzzle would eliminate the danger of hood cave-in when the gun was fired immediately over the engine.

Right rear view, Second pilot

Detail of the tandem suspension

Most of the recommended changes were incorporated in the second pilot before it was delivered to Aberdeen in March 1942, after a drive of 813 miles which consumed 38½ gallons of fuel. Further modifications were made at the Proving Ground, including the installation of a 20-gallon fuel tank and reduction of the height of the gun shield. There was little resemblance between the shields of the first and second pilots; the second, though mounted lower, was shaped to a point above the gun sight. It was cut off and squared, at approximately the same level as the horizontal seam of the first pilot shield, resulting in a lower silhouette than that of the first pilot. The height of the shield itself was later reduced from 35⅝ inches to 21⅝ inches, and the over-all height, excluding gun sight, from 72¼ inches to 58¼ inches.

Second pilot, front right view

The second pilot was given complete automotive tests at Aberdeen, first equipped with 6.00 x 16 tyres and later with 7.00 x 15 tyres and wheel equipment. It was also given a firing test in which it compared favorably with the T8.

At the conclusion of the tests, it was reported that the 37mm Gun Motor Carriage T14 was mechanically suitable for its intended purpose because of its mobility, low silhouette, mechanical simplicity, ease of maintenance, and small size. Besides these qualities, the following advantages were cited: excellent riding characteristics; adequate power and traction in mud and sand operations, with flotation increased by the larger (7.00 x 15) tires; excellent cooling properties; satisfactory firing stability when the vehicle was stationary; good slope-climbing ability; and satisfactory visibility, approximately 15 feet ahead of the driver's seat. About 60% of the parts were interchangeable with parts of the standard ¼-ton 4x4 reconnaissance car, in this case the Willys "Jeep". There were still some downsides, however. For example, with the two ammunition frames in place, the gun would only traverse 30 degrees to either side. But if the third item on the deficiency list was that the horn button was inconveniently placed behind the gear lever, it probably puts the majority of the deficiencies in perspective.

Front left three-quarter view, Second pilot

By this time, the M6 was already in quantity production and it was considered inadvisable to standardize another 37mm gun motor carriage. There was, however, a need for a small, lightly armoured vehicle for use in reconnaissance, command, security, and messenger service in forward areas. A letter from the Tank Destroyer Headquarters, Tactical and Firing Center, Temple, Texas, to the Adjutant General recommended that a project be initiated to develop such a vehicle on the T14 chassis. Accordingly, the second pilot, after undergoing service tests at Fort Bragg, N. C., was used as a basis for Scout Car T24. The 37mm gun was removed and the vehicle, with armour, was tested by the Tank Destroyer Board and the Armored Vehicle (Palmer) Board at Aberdeen Proving Ground. Both boards recommended its standardization.

However, the Commanding General, Army Ground Forces, considered that the vehicle was not suitable for any of the using arms except the Tank Destroyer Command, whose tactics made it unnecessary for all personnel to ride behind armour. The Development Branch, Services of Supply, concurred in this view, and the project for Scout Car T24 was terminated by Ordnance Committee action in January 1943. The pilot vehicle was returned to Aberdeen Proving Ground for use as a proof facility.

Right side view, 37mm Gun Motor Carriage T14, without windshield

There was no further development of the project for a 37mm gun motor carriage on this chassis. The first pilot of 37mm Gun Motor Carriage T14, with modifications, was shipped to Holabird Quartermaster Depot for test as a general purpose ¾-ton vehicle and also to Wright Field, Dayton, Ohio, for tests by the Air Corps. Some were modified into tractors to haul 2-ton trailers. One new-build was delivered to the Medical Service Corps as an ambulance, and found acceptable. As a general overview, though, it was found to have a number of deficiencies compared to the then-current ¾-ton 4x4 truck, in areas such as turning circle, ground clearance, and fordability, and was not recommended for production. Development of the basic vehicle, the Willys ¾-ton 6x6 truck, was canceled by Ordnance Committee action (OCM 20641) on 21 May 1943.

Side view, 37mm Gun Motor Carriage T14, second pilot with windshield

Above and left: Views of the second pilot vehicle after modifications made to lower the profile of the gunshield. Photos taken April 1942.

Above: T14, second pilot, front view 20 March 1942, Aberdeen Proving Grounds

Above left side view, 37mm Gun Motor Carriage T14, second pilot

37mm Gun Motor Carriage T33

Above: 37mm GMC T33 in final configuration
Right: The Cargo carrier before converison

On 22 October 1941, a letter was sent to the Chief of Ordnance by Aberdeen Proving Ground:

On Septemper 22, 1941, the Ford ¾-ton 4x4 Cargo Carrier came to Aberdeen Proving Ground for a very brief visit.

In view of the favor with which the 37mm Gun Motor Carriage T-21 was received by the troops it is thought desireable to further

exploit this vehicle. This vehicle will give much greater visibility for the driver than the T-21, and with the addition of a small windshield would have all the advantages of the T-21. This vehicle has the same type body which is on the T-21 and would be capable of all-around fire. As can be seen [in photographs] this vehicle has a lower silhouette and the driver's seat is lower than in the T-21.

It is recommended that this vehicle be procured for experimental use as a gun motor carriage

The response came on 18 November, with an order that the vehicle be built up as a 37mm Gun Motor Carriage for comparative test with the T21, and that the 37mm M3 be mounted in a similar manner to that of T21. Except that it had a front-mounted engine and forward driving position, engine alongside driver, the chassis was similar to that of the 37mm Gun Motor Carriage T8, based on the Ford "Swamp Buggy." The new vehicle was designated 37mm Gun Motor Carriage T33.

The primary modification at Aberdeen Proving Ground was the removal of the cargo body and its replacement with a welded gun pedestal and base, bolted in place on top of the frame. A 37mm Gun M3 was mounted with shield and traveling lock similar to the type used on the 37mm Gun Motor Carriage T21. The running boards were replaced by ammunition compartments. Two seats facing to the front were mounted on the rear of the vehicle, and one seat, facing to the rear, was placed behind the driver's seat. Safety belts were provided for all crew members.

Limited testing of the vehicle was conducted between 21 November 1941 through 15 January.

Above and below: The T33 part-way through the conversion process. Bottom: T33 complete

Above: Two more views of T33. On the left, demonstrating the traverse to the rear
Right: Demonstration of the operation of the gun travel lock

Conclusions were good.

The subject vehicle is a satisfactory motor carriage for the 37mm Anti-Tank Gun and is superior to previous mounts of this type for the following reasons:

a. Lower silhouette - present height is 80" but can be reduced to 75" by including the pedestal base in the vehicle frame and lowering the rear fenders.

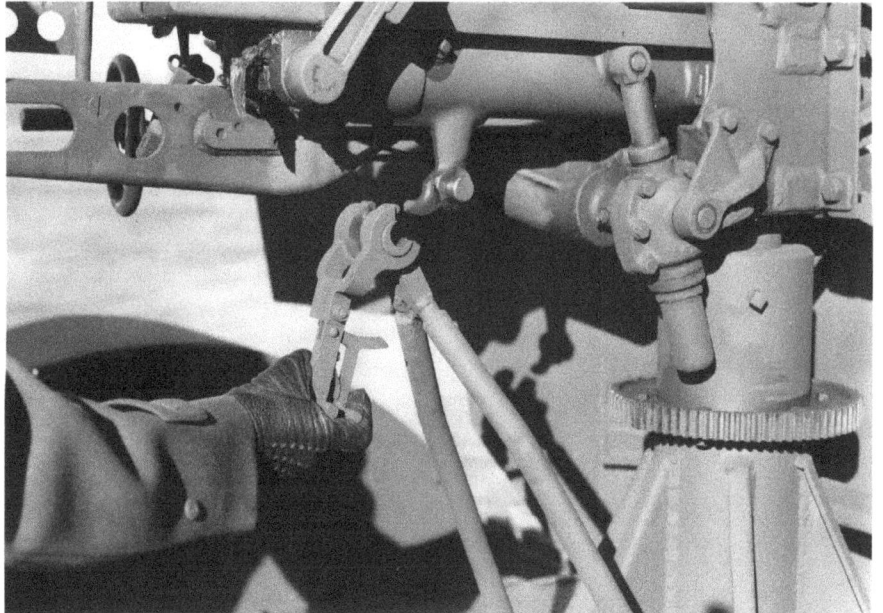

b. Lighter and more rigid pedestal mount - approximately 100 lb. lighter than pedestal and base used on 37mm Gun Motor Carriages T-8 and T-21

c. Greater range of fire - the gun can be fired at elevations of -10° to +15° at all traverse positions except for a small sector to the front where elevation is limited to 0°to +15°. When firing to the front, the windshield will be cracked but instruments undamaged. (A windshield of shatterproof plastic could be tried for this purpose.) This performance is equivalent to that of the 37 mm Gun Motor Carriage T-8 and is superior to that of the 37mm Gun Motor Carriage T-21 which can only be fired at maximum elevation to the front to avoid damaging the instruments.

d. Better driver visibility - the limit of vision of the driver to the front of the vehicle is 15 feet measured from the driver. The comparable measurement on the 37mm Gun Motor Carriages T-8 and T-21 is 20 feet.

e. Performance characteristics at least equal to those of the above mentioned gun motor carriages.

Right: There are always casualties in testing. This windshield laid down its life for the cause

In a teletype to the Commanding General at the Proving Ground, on 16 January, the Ordnance Department requested that a limited automotive test be made of modifications incorporated in the T33, after which the vehicle be held inactive pending the result of studies, then in process, of designs for mounting a 6-pounder gun. The 37mm gun was proof fired, and the mount declared adequate but no more formal testing was conducted.

Left: The steering wheel swung away to allow easier access. Right: 37mm GMC T33, front view

The Aberdeen report recommended that the vehicle be adopted as a standard 37mm gun motor carriage, and that the pedestal and base used for mounting the gun be adopted for use on all 37mm gun motor carriages of this type. It was further recommended that on future models the pedestal base be incorporated into the frame of the vehicle; that the rear fenders be lowered to the minimum of clearance; that the traversing mechanism be replaced by the inclosed gear type of mechanism in use on the 37mm Gun Motor Carriage T21; and that a shatterproof plastic windshield be considered for use on future models.

In the meantime, however, comparative tests had been made between the 37mm Gun Motor Carriages T21 and T8, and the T21 had been standardized as the M4 (later the M6). Because of the similarity of the T33 to the T8, and the fact that it was finished after these comparison tests were made, the T33 was never subjected to a service test by the using arms.

In April 1942 a letter from the Services of Supply stated that cancellation of the T33 was desired. The 37mm Gun Motor Carriage M6, based on the Standard Quartermaster ¾-ton 4x4 chassis, was already in production, and it was directed that the T33 design be dropped in its favor. The project was closed formally by Ordnance Committee action; the vehicle was disassembled, and the chassis was used as a basis for the 57mm Gun Motor Carriage T44.

Left side view, 37mm GMC T33

37mm Gun Motor Carriages T22, T23, T22E1, T23E1

Above: 37mm GMC T22 with turret to the rear and drivers' armoured panels in raised position

A memorandum from the Assistant Chief of Staff, G-3 (Operations), to the Assistant Chief of Staff, G-4 (Supply), dated 30 July 1941, submitted military characteristics for an all-purpose vehicle, capable of mass production at minimum cost, to be used primarily as a tank destroyer. The vehicle was to be light, highly mobile, lightly armored, with low silhouette, and was to mount a 37mm gun.

Resulting design studies indicated that a vehicle meeting most of the proposed characteristics could be manufactured on a mass production basis, using a 6x6 chassis. The basic design was to be adaptable for use as a mobile mount for a 37mm gun and also for other weapons such as ground machine guns, dual or quadruple cal. .50 machine guns, dual 20mm antiaircraft machine guns, or an 81mm mortar. It was also intended to be usable as a cargo carrier for battlefield ammunition supply.

Ordnance Committee action in October 1941 recommended the procurement of pilots with the proposed characteristics from two manufacturers, in order to have tests of two different designs and thus determine the more satisfactory type. Procurement of eight pilots was authorized in December by the Adjutant General, and the following 37mm gun motor carriage designations were assigned: T22 (6x6,Ford Motor Co.); T22E1 (4x4, Ford Motor Co.); T23 (6x6 Fargo Motor Corp.); and T23E1 (4x4, Fargo Motor Corp.).

Top: T22 front view. Bottom: Turret interior

At this time, a need existed in the Tank Destroyer Battalion and the Cavalry for several types of light armored cars; the T22 and T23 vehicles, with slight modifications, were believed to fill the requirements. They were redesignated, therefore, while under procurement, as Light Armored Cars T22, T22E1, T23, and T23E1.

The first pilot of Light Armored Car T22, after being demonstrated at Aberdeen Proving Ground and at Fort Knox, Ky., was recommended for standardization, with numerous modifications, by the Armored Force Board. This vehicle, the 6x6 type built by the Ford Motor Co., was modified, as T22E2, and adopted as Light Armored Car M8.

In the meantime, T23 seemed to be finding more favour with the Tank Destroyer Board. Initial investigations indicated that T23 could not mount the 57mm cannon, though Fargo did express the opinion that they'd be willing to try to convert the vehicle to mount it. A Tank Destroyer Board observer commented, however, that to do so would likely require a pedestal mount, and even at that lengthening the vehicle by at least 8". The alternative was to create a 60" turret ring instead of the extant 48", but since the weight of gun and ammunition would be excessive, some 1,000lbs of armour would have to be shaved off to keep within performance acceptable.

Left, two views T22. Below, T23

Above: T23, left side and rear. These photographs, and those of T23E1 are from the operators' manual, evidently written before the addition of armament.

The observer also commented most favourably on the stowage space in comparison with that available on T22, on which he opined that it would be difficult to fit the recommended equipment issue.

On 29th January a delegation from Chrysler Corporation consisting of Messrs Dalton, Maddox and Fisher arrived at Ft George G Meade, Md, for a meeting with the Tank Destroyer Board. During this meeting, the following suggestions were made:

a. The need for the design of a super-structure to adapt the T-23 and T-23E1 gun motor carriages to the needs of the tank destroyer units. The design suggested, in effect, is the T-23 and the T-23E1, without turrets, with light armor and with the top open to approximately 82" x 52". The interior to be provided with bench seats folding down or up. A skate mounting to be provided on the well opening for mounting a .50 or .30 calibre machine gun either singly or in pairs. This vehicle to be used as an all-purpose carrier for troops or ammunition, as a reconnaissance car, or as a gun carriage for the 37mm or 57mm guns. These guns to be mounted on a pedestal.

b. The importance of providing armor on the Fargo truck to give protection against small arms fire to the engine, the rear and front axle housing, and the driver.

c. The possibility of extending the winch axle the entire width of the vehicle for a pair of power-driven wheels. These wheels to be approximately 15" in diameter with a 1" tread, to be made of metal with a rough facing. The purpose of these wheels is, when the vehicle is bowed down to apply the power and pull itself out on its own power. With this attachment the wheeled vehicle should be capable of accomplishing all that the tracked vehicle accomplishes in rough terrain.

T23E1

Whether those suggestions were ever acted upon is information the author has not been able to confirm. A telephone call on 30 March between Mr Biggers and General Bruce indicated that there was still, even after the reclassification of the vehicles from motor carriages to armored cars, interest in T23 from the Tank Destroyer Board. Tank Destroyer Branch were still looking at T23 for the gun carriage role in addition to several other roles in TD Branch. General Bruce proposed again putting the 57mm into the vehicle, with the full understanding that it would have to have the turret "half lopped off". He considered that about 45° traverse "would be OK." A shorter 57mm was being considered by Ordnance with a shorter recoil. "We are willing to cut off the turret; anything as an expedient"

Mr Biggers replied that he wasn't sure the vehicle would take the 57mm recoil regardless, but that he would look into it.

Above: T23E1, front and rear. Below: T23E1 driver's compartment, and turret interior

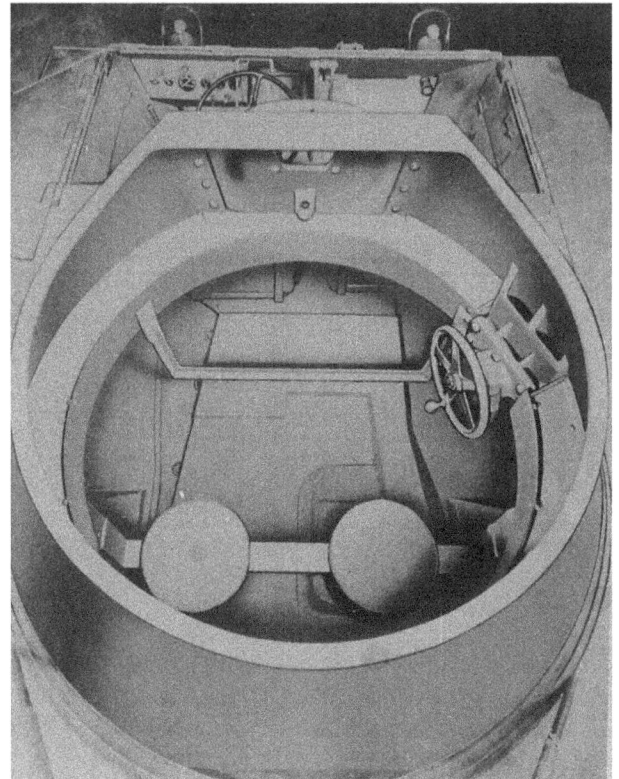

If he did actually do so, it was for little gain. In view of the standardization of the T22E2, the projects for Light Armored Cars T22E1, T23, and T23E1 were closed by Ordnance Committee action in April 1942. The pilot vehicles were completed for use as proof facilities.

37mm Gun Motor Carriage T43

On 29 January 1942 the Studebaker Corporation offered to build and submit for test, without charge to the Government, a pilot vehicle complying with the military characteristics approved for the 37mm Gun Motor Carriages T22, T22E1, T23, and T23E1. The offer was accepted and the pilot was designated 37mm Gun Motor Carriage T43. OCM 17929 of 12 March 1942 reclassified and redesignated the vehicle as Light Armored Car T21. Thus ended the TD Board interest in GMC T43

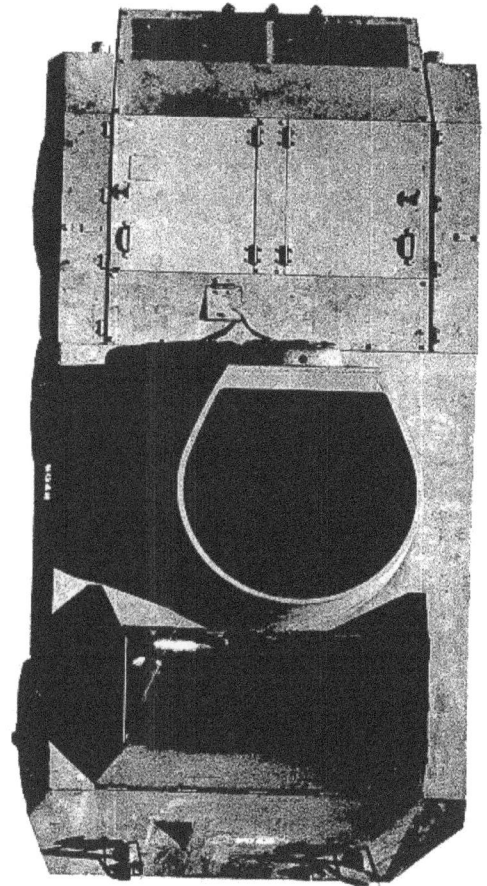

Above and Below: 37mm GMC T43 after redesignation to Light Armored Car
Above right: Top view of T23E1, photo didn't fit in the previous entry!

Scout Car M3A1E3 with 37mm Gun

Scout Car M3A1E3 with 37mm Gun on Mount T6 in its original configuration showing maximum depression shortly after arriving in Camp Funston, KS, photo taken November 1941

On 4 June 1940, the Chief of Cavalry recommended the mounting of the 37mm Anti-Tank Gun M3 on a Scout Car M3A1, to provide 360° traverse with the elevation and depression as provided on the standard M4 ground mount. He made available both the car and the gun. The pedestal would be located near the front of the car body to provide for depression with the gun firing forward. It was considered that considerable stiffening of the body structure would be required, and the two .30 machine guns on skate mounts would be retained. In accordance with OCM 15865, Wellman Engineering Company modified the M3A1 by adding Gun Mount T6, the vehicle was then driven to Aberdeen Proving Ground for testing between 9 July 1941 and 3 August 1941. Test observations were as follows:

Noise Observation: This feature of the Scout Car M3A1E3 is very good. Being a wheeled vehicle it can be operated as quietly as any motor vehicle of its size.

Visual Observation: The Scout Car itself has a low silhouette and is comparatively easy to conceal. The T-6 mount, however, increases the overall height of the vehicle to 96-3/4 inches, but the increased height is only the gun itself, which does not add a great problem.

Left: The original mount in Aberdeen

The vehicle was driven over the washboard course and no interference was noted. As far as can be noted the mounting has not caused any changes in the suspension flexibility or the riding qualities of the vehicle. No difficulty was experienced with loading or manipulating the controls of the gun. It is believed, however, that the rate of aimed fire could have been materially increased if a semi-automatic breech mechanism such as that of the M6 gun had been used.

The comfort and safety of the crew is no way different than that of Scout Car M3A1. It was thought that muzzle blast would be injurious to occupants of the driving compartment, but by actual firing tests no discomfort was noted.

Not all was good, however. The conclusion was:

The Scout Car M3A1 is too large a vehicle to be employed merely as a gun motor carriage for the 37mm gun.

It is recommended that the Scout Car M3A1 should not be used as a motor gun carriage for the 37mm anti-tank gun.

Upon the completion of the tests, the vehicle was sent to Camp Funston, Kansas, where it was tested by the Cavalry Board. Upon arrival, the Cavalry displaced the gun mount by rotating it 180 degrees, and moving it some 17½" to the rear, with a view to obtaining 360 degree traverse. Combined with the offset at the top of the mount, the entire gun was displaced 23½"

Scout car M3A1E3 with 37mm gun on mount T-6 (Modified), February 1942

The downside was that the displacement of the mount resulted in the muzzle resting over the driver's head. It was therefore necessary to fabricate a muzzle blast device to protect him and the instrument panel. (They called it a "Muzzle Blast Arrestor"). This extended the overall length of the weapon by 18¼."

In a report dated February 16, 1942 it recommended that a more satisfactory gun mount for the 37mm gun in the Scout Car should be designed; namely, a mount which would permit more freedom of movement of the gun crew. It was suggested that the gun be displaced 23½" to the rear of the vehicle; that the diagonal braces be eliminated from the mount; that a free-traverse feature for the gun be incorporated into the mount; that a muzzle blast arrester be added to the gun and that the gun be equipped with a tank gun sight such as the telescope M5A1. It was decided that there was a definite Cavalry need for a 37mm self-propelled mount and that the M3A1E3 was a satisfactory provisional carrier for the 37mm gun mount.

As a result, the vehicle was then returned to Aberdeen Proving Grounds for the replacement of the mount. A letter from General Barnes recommended that the 37mm Gun Pedestal Mount M25 as used in the 37mm GMC M6 be used with minor modifications, using the fitted Telescope M6. The letter did opine, however, that "the use of the 37mm Gun Motor Carriage M6 offers the best solution to the problem. This motor carriage is a standard vehicle now in production"

The modified mount showing its ability to be serviced facing to the rear

A hollow steel box structure was constructed to bolt to the two longitudinal frame members of the vehicle and to carry the 37mm Gun Mount M25 high enough so that the 37mm gun would be in the position recommended.

Other modifications to the vehicle consisted of modifying the floor plate to fit around the gun base, cutting out the forward

Close-up of the reversed mount and with the blast arrestor fitted

bulkhead to provide more space for the gun crew to operate, and providing a hinged cover for the rear floor well so that a better firing platform would be obtained. The subject vehicle with the gun mounted was viewed by a Cavalry representative and was decided to be generally satisfactory. Numerous suggestions were made by the Cavalry representatives at that time consisting mostly of requests for the installation of several radio sets.

The vehicle was subjected to brief testing after installation of the M25 pedestal. It was concluded that "the subject vehicle constructed at Aberdeen Proving Ground fulfills the requirements as set forth in the original Ordnance Office directive."

Scout Car M3A1E3 with 37mm Gun on Mount M25

However,

The application of the 37mm Gun Mount M25 to the Scout Car M3A1E3 results in a vehicle of ungainly appearance and high silhouette which is at best an expedient.

The 37mm Gun Motor Carriage M6 is a superior vehicle in all respects, except armor protection, to the Scout Car M3A1E3 with 37mm Gun Mount.

Thus it was recommended that:

If the Using Arm has a definite need for a 37mm gun mounted on the Scout Car M3A1, the subject vehicle be considered for use.

Since the 37 mm Gun Motor Carriage M6 is a superior vehicle for carrying the 37mm gun, any requirements for the Scout Car with 37 mm gun mount be changed to provide for its use.

Add in the fact that the 37mm was by this point rapidly becoming obsolete, and the program, never officially given Gun Motor Carriage project status, was dropped shortly thereafter.

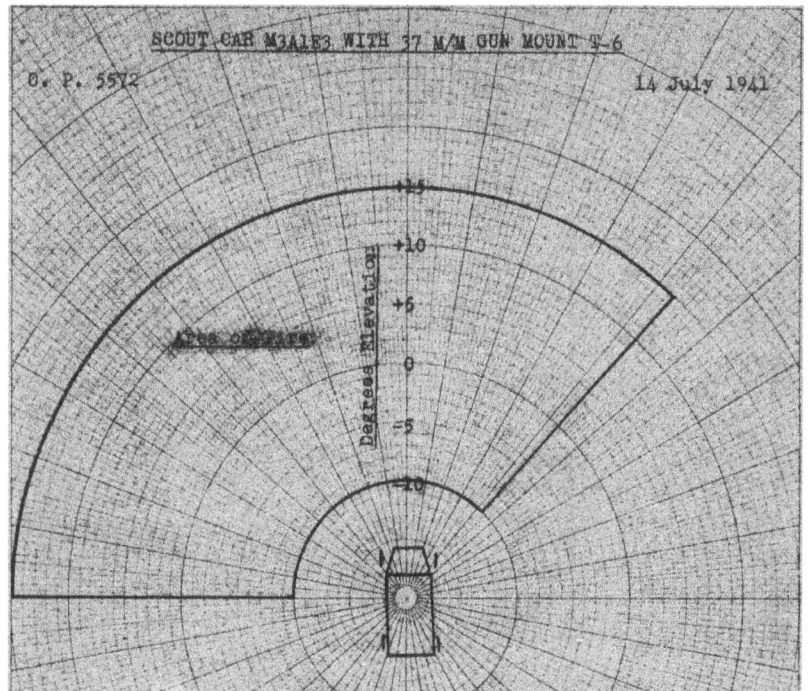

In comparison with the 360 degree arc of fire the vehicle ended up with, this was the original arc.

57mm

Several projects for the development of gun motor carriages mounting the 57mm Gun M1 were initiated, but none of the vehicles were standardized by the US.

The gun had been developed because British battle experience indicated the need for a light gun more powerful than the 37mm gun or 2-pounder for antitank use. This requirement was met by the British 6-pounder gun, which, converted to U.S. gears, threads, and tolerances, was standardized in 1941 as the 57mm Gun M1. Mounted on Carriage M1, it could be depressed to -5°, elevated to 15° and could be traversed 45° right and 45° left. Fired from this gun, the APC projectile M86 had a muzzle velocity of 2,700 feet per second and a maximum range of 13,555 yards. It would penetrate 2.8 inches of homogeneous armour plate at 30 degrees obliquity at 1,000 yards.

However, the carriages were soon overtaken by events. Because of the irresistible trend in World War II toward more powerful guns and of the success in developing chassis mounting these guns, the development of 57mm gun motor carriages was discontinued. Further, as will be seen in the entry for 57mm GMC T49, the 57mm was not seen as an improvement overall for TD use over the 75mm anyway.

57mm Gun Motor Carriage T44

57mm Gun Motor Carriage T44 with gun travel lock engaged

The idea of mounting a 57mm cannon on the T33 chassis in place of the 37mm was first mooted in a telephone call on 26 January 1942 between a LT Keeler and a CPT Klima. By 29 January it was requested that as soon as test on T33 were completed, the gun be dismounted and replaced with a 57mm. A proposed drawing was attached to a letter from Office Chief of Ordnance to Aberdeen Proving Ground, which gave only the position of the gun on the chassis: Aberdeen was tasked with actually designing and creating the mount. A minimum crew of three was desired, four if possible. Initial investigations with Ford had indicated that the vehicle could be loaded to up to 7,000lbs, and if the design worked, heavier duty axles could be provided.

Mount showing the maximum depression.

Given the knowledge that the vehicle would be overloaded, the tests which Ordnance Dept requested were purely related to the gun and mount. They specifically directed that no automotive tests be conducted with the exception of cross-country tests used to determine the effect of the gun pedestal and supporting structure. Ordnance directed that firing testing should be done to determine the suitability of the chassis as a gun platform, with an eye to the following produced data:

a. Weight of gun, pedestal, and base structure and estimate of weight saving possible by redesign of pedestal and base.
b. Area of platform required to adequately serve the gun.
c. Degree of lock-out of flexible vehicle suspension members to give adequate stability to the gun platform.
d. Plan of chassis arrangement and location of seats, ammunition boxes, etc., found to be desirable.
e. Any other factors which should be considered in the redesign of the basic chassis to provide an adequate Gun Motor Carrage.
f. A weight estimate of an adequate gun shield similar to those designed for the 37mm Gun Motor Carriages, T21 and T33.

Above: T44 part-way through the conversion process, at maximum depression. Note the compression of the springs caused by the weight of the larger gun even before the 300lb shield was added. Compare with the equivalent photograph of T33 some pages back.

The gun pedestal and base was designed similarly to that used on the 37mm GMC T33 and was based on a calculated average trunnion reaction of the 57mm gun of 4,900lbs. The travel lock used was also similar to that on T33. The mount retained the traverse of 45 degrees to each side, and -10° +15° range of elevation of of the towed carriage.

It was decided that the most practical method of carrying ammunition would be in boxes carried on both sides of the frame between the front and rear wheels. The racks which were constructed carried 56 rounds in the vertical position. In order to secure a maximum of protection for the gun crew, an angular shield of ¼" plate was constructed. The vehicle proved too small to allow adequate seating for the three-man crew and so this detail was omitted.

49

By 16 February 1942, the pilot model was ready for testing at Aberdeen. However, due to the lack of a certified 57mm gun, firing tests did not occur until March 11. It proved a short testing process.

A total of seventeen rounds was fired at center and maximum right and left traverse and at center, maximum and minimum elevations, both with and without the application of foot brakes. Without brakes applied, recoil forces would push the vehicle some three feet foward. With the gun at centre traverse and with foot brakes applied, performance was reasonably satisfactory, the carriage suffering a net displacement of only two or three inches. When the gun was fired at maximum right traverse, however, the left rear wheel left the ground approximately 18 inches and great pitching of the carriage occurred.

Above: The left side ammunition stowage bins
Right: 57mm GMC T44, rear view

The proof officer was a 2LT Depkin. The seventeen rounds were enough for him, and he stopped the test. His report commented:

The overturning moment of the 57mm gun is entirely too great to allow its mounting on a light vehicle of this type.
The type of transmission hand-brake installed on the subject vehicle is inadequate for holding the gun motor carriage in position during firing.
When the gun is operated semi-automatic, the cases are ejected at such a velocity as to constitute a danger to driver and crew.

The Proving Ground's final report also concluded:

The vehicle was too light to carry the required load of approximately 3,500 pounds.
The firing platform on the subject vehicle was adequately wide, but not sufficiently long to allow proper servicing of the gun or seating for the crew members.
A three-man crew would not be adequate to properly serve the gun and drive the vehicle because of the weight of the projectiles which must be handled and the necessity for replacement of casualties.
The 57-mm Gun M1 when adapted to a self-propelled mount could, with very slight modification, be provided with 360° traverse. Although firing from the broadside position would cause extremely unstable action of the carriage, in an emergency the additional traversing ability would be of great assistance.

Above: 57mm GMC T44 at maximum elevation
Right: Front view

When the results of the tests reached the Office, Chief of Ordnance, a teletype was immediately sent to Aberdeen to suspend all further work on this mount and to hold the gun and carriage in abeyance until a more suitable mounting could be designed.

On 20 April 1942, General Barnes sent a memorandum to the Secretary of the Ornance Committee:
This vehicle was constructed for the purpose of securing experimental data. Results indicated the unsuitability of the ¾ ton 4x4 chassis as a mount for the 57mm Gun M1 and the necessity of using a larger, heavier vehicle for mounting this gun. In view of the above, this development is considered completed and the project is closed.

Top: The mount in its maximum right and left traverse, also maximum elevation. Above, below: Three-quarter views of the complete vehicle

The 57mm Gun M1 tested on this gun motor carriage was used in the development of the 57mm Gun Motor Carriage T48. The 4x4 chassis was reassembled with its original cargo body and returned to the Quartermaster Depot from which it had been borrowed. The project was formally closed by the Ordnance Committee in April 1942. This was the only official attempt known to the author of a wheeled 57mm motor carriage.

75mm

Although not specifically considered as a dedicated anti-tank gun, the 75mm was pressed into service as an expedient in the role, particularly in the M3 half-track. However, various attempts were made to mount a gun of such calibre onto a wheeled vehicle.

75mm Gun Motor Carriage T27

75mm Gun Motor Carriage T27, left side.

On 21st August 1941, Colonel Christmas sent to Aberdeen a letter. Extracts follow:

The Studebaker Corporation have recently submitted to Holabird Quartermaster Depot a 1½ Ton, 4 x 4, Low Silhouette, Chassis for test to determine its suitability as a vehicle within the 1½ ton classification of the proposed new 4x4 short wheel base, low silhouette, Quartermaster vehicles. The vehicle now incorporates four-wheel steer and two-way drive; neither characteristic being a requirement under the Quartermaster program.

The Ordnance Department has been directed to cooperate with the Quartermaster Corps in their studies of the proposed line of ¾ ton, 1½ ton, and 3 ton vehicles with a view to determining the suitability of these chassis as gun motor carriages. With this view in mind, the Quartermaster Corps have offered to make the subject vehicle available to the Proving Ground for approximately two weeks to determine the suitability of vehicles of this class as motor carriages for the 75mm guns. Preliminary studies of the vehicle and drawings have indicated the feasibility of mounting and firing the 75 mm gun from this chassis.

It is requested that upon receipt, the 75mm gun used on the pilot 75mm Gun Motor Carriage, T12, be mounted upon the Studebaker chassis and sufficient firing and cross country operation be conducted to determine the suitability of this class of vehicle as a motor carriage for the 75 mm gun. Necessary modifications may be made to the vehicle proper.

By the time the test report was written, the official object of the test was "To determine the relative stability of a wheeled vehicle in comparison with a half-track vehicle when used as a mount for a 75mm gun." The short test window provides an interesting overview of the process if one is in a hurry.

Two vehicles arrived from Holabird on 6th October 1941. They were of slightly different configuration. The vehicle designated #1 had four-wheel steering, and had a rear-mounted engine. Vehicle #2, on the other hand, was equipped with dual wheels at each corner, with the engine mounted amidships on the right side.

An initial test of the two vehicles was conducted that day. It was found that the feature providing 4-wheel steering on the #1 vehicle led to general instability and difficult handling, due largely to the low steering ratio and lack of provision for self-righting of the wheels. This instability was not so pronounced on the #2 vehicle and the general handling characteristics of this vehicle were very satisfactory and stability in turns was excellent. As a result, vehicle #2 was selected as the ideal candidate for the gun.

Two views of Vehicle #2

7 October saw the removal of the gun and mount from GMC T12 in preparation for installation on vehicle #2. However, initial difficulties dictated that in fact vehicle #1 would be the first to receive the gun, work on this starting on 8 October. This was not completed until 15 October. In the meantime, work progressed on vehicle #2 starting 11 October. This job was never completed before it was time to return the vehicles.

After photographing, the vehicle was test driven and fired on the 18th October. In the interests of expediency, no attempt was made to re-attach pieces of the vehicle removed to make room for the mount. The mount itself had to be modified to be approximately 2½ inches higher to clear the transfer case.

The vehicle was negotiated 20 times over the washboard course at the maximum practical speed (10mph - limited by ability of driver to stay with the vehicle) without any measurable permanent distortion to the frame.

Vehicle #1 stripped down as part of the conversion process

The gun was fired at various positions of maximum elevation, traverse, and depression with charges up to 115% pressure.

Right side and front views of the 75mm GMC T27

The report concluded as follows:

1. That the Studebaker mount is not as stable as the HalfTrack Gun Motor Carriage T12 when used with the 75 mm. gun.
2. That the vehicle would be overloaded with the 75mm gun, crew, and ammunition.
3. That the handling characteristics of the Studebaker #2 chassis are superior to those of the #1 chassis.
4. That the frame was adequate for the installation as tested.
5. That the #2 Studebaker vehicle would probably make a very suitable mount for the 37mm anti-tank gun.
6. That at least 180° fire would be desirable from the vehicle but that this degree of traverse could not be safely attained with the vehicle and gun tested.

It was thus recommended:
1. That no further consideration be given the number 1 Studebaker chassis as a mount for the 75mm gun;
2. That a larger wheeled vehicle is necessary to provide a stable firing platform for the 75-mm gun;
3. That the #2 Studebaker be considered as a motor mount for the 37-mm. antitank gun;
4. That if the #2 chassis is considered as a 37mm antitank mount, a self-righting steering device be incorporated in the present mechanism.

The two-week period having expired, the gun was removed and the chassis returned to Holabird Quartermaster Depot. The project was closed by Ordnance Committee action on 20 April 1942. (OCM 18125)

75mm Gun Motor Carriage T66

75mm GMC T66 on 05 February 1943

Following the manufacture of the first pilot Armored Car T19 and the decision to build a modified vehicle designated Armored Car T19E1 on 01October 1942 (OCM 18962), a project was initiated to mount a 75mm Gun M3 on an Armored Car T19E1 for consideration as a gun motor carriage as an expansion of the T19 project.

The Ordnance Committee, on 10 October 1942 (OCM 19121), officially gave military characteristics for the proposed vehicle and recommended that a pilot model be built and designated 75mm Gun Motor Carriage T66. What is of particular note, however, is that the T19 project was originally a project of the Tank Development Branch. An internal memorandum dated 04 August 1942, before Ordnance Committee got particularly involved, indicates that Chevrolet had been already instructed to create a version of the T19 using a 57mm gun, furnishing everything including the turret but the gun and mount. That same memorandum also suggests that Chevrolet be given written confirmation of the change to the 75mm which had been verbally instructed to them previously, with an additional note to let the Ordnance Committee know so that they could 'read it into the record.'

The Development Branch assumed supervision of the project and redesigned the pilot by replacing the GM truck engines with Cadillac V8s with hydramatic transmissions, enlarging the hull, substituting 14.00x20 tyres for the 12.00x20 tyres used, and by generally improving the hull design. Particular attention was paid to the reduction of weight. Armored Car T19E1 was to be held to a fighting weight of 28,000 pounds and the 75mm GMC to about 30,000 pounds. Individual propeller shafts terminated in a transfer case which divided the torque of the two engines by a planetary differential with lock-out feature, ⅓ to the front differential and ⅔ to the rear. The ⅔ torque was divided by another differential with manual lock-out between the rear differentials.

Above, a rear view of T66 in its original configuration before the modification of the turret

Below, a detail view of the individual coil sheel suspension with the inside shock absorber at full extension.

Thus the three axles equally spaced each received ⅓ of the power and were provided with a means of manually locking out the differential action between them. Suspension was of the independent swing arm type with helical springs operating around a shock absorber. Approximately 13" of vertical motion was obtainable.

As to the crew, driver and assistant driver were conventionally located with hatches directly overhead. Both turrets had places for three crew members, commander, gunner and loader, T66's turret being open-topped. The pilot 75mm Gun Motor Carriage T66 , which was manufactured concurrently with Armored Car T19E1, utilized the same hull and chassis but mounted a 75mm Gun M3 in a modified Combination Gun Mount M34 in a semi-open turret. A similar turret was used in 75mm Gun Motor Carriage T67, which was being developed at the same time, in order to compare wheeled vehicles and track-laying vehicles for use as gun motor carriages.

In November 1942, while Armored Car T19E1 and 75mm Gun Motor Carriage T66 were under construction, the Armored Vehicle Board met at Aberdeen Proving Ground to review the general program for armored cars and gun motor carriages. In its preliminary report, which appeared in December 1942, the Board recommended that further development of Armored Cars T19 and T19E1 be terminated. It also stated that it did not consider the T19E1 suitable for development as a gun motor carriage for the Tank Destroyer Command.

In view of this report and its approval by Headquarters, Services of Supply, the Ordnance Committee recommended that the development projects for Armored Cars T19 and T19E1 and for 75-mm Gun Motor Carriage T66 be closed but that the pilot models of the T19E1 and the T66 be completed and given engineering tests for purposes of record since the contract with Chevrolet had been agreed anyway, and it was the most desirable way of ending the contract.

The two pilots were completed and shipped to Aberdeen Proving Ground for testing. 75mm Gun Motor Carriage T66 was to have been given an accelerated road test and proof firing of the gun similar to the Armored Vehicle Board test. The gun was proof fired, but the vehicle did not have a thorough road test because the planetary gears in the transfer case broke one hour after the test began, and replacements were not available in time. Aberdeen reported that this vehicle had riding qualities superior to those of the current track-laying vehicles and that its speed on concrete or good secondary roads was 57 m.p.h., superior to that of current tanks. The vehicle was considered a satisfactory

A frontal view of T19E1

mount for the 75mm gun, even though considerable movement during firing was found. The Proving Ground reported, however, that the T66 failed to fulfill one of the major requirements of a gun motor carriage: Namely to have cross-country mobility fully equal to that of a tank. It was also stated that the special two-speed transfer case needed further development.

Aberdeen Proving Ground recommended, therefore, that further development work on the vehicle be terminated and that future development of gun motor carriages be directed toward the producing of tracklaying vehicles with superior cross-country mobility.

Following the tests at Aberdeen, Gun Motor Carriage T66 and Armored Car T19E1 were sent for testing to the Desert Warfare Board at Camp Seeley, California, where they were to be compared in overall performance, including flotation, with half-track cars, the 3" GMC T55, and light and medium tanks. After these tests, which indicated that the six-wheeled vehicle had excellent riding qualities and flotation, Armored Car T19E1 was sent at the request of the British to Camp Young, California, for comparative testing with Armored. Car T17E1, but the 75mm Gun Motor Carriage T66 was retained at Camp Seeley for use as a facility vehicle.

Top: The engine compartment, looking to the rear.
Above:The suspension, under load, with drag link and pitman arm visible.

Subsequent operation of these two pilots resulted in failures to the point where sufficient components remained to maintain only one complete operating vehicle. Because the T66 represented a more unique design, it was retained at Aberdeen Proving Ground for historical purposes, and Armored Car T19E1 was scrapped. This was ordered by OCM22754 of 27 January 1944.

Left: Centre axle at full extension
Right: Interior of T66 fighting compartment

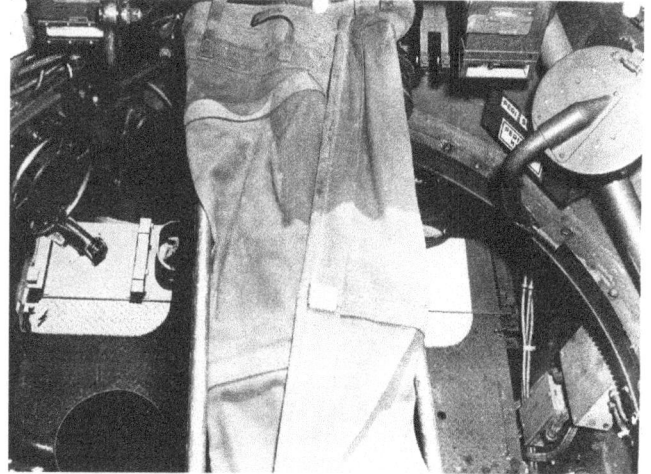

There is, however, an unknown epilogue to this story. T66 was at some point given additional armour plating to the turret and is photographed in testing with the photographs dated April 1944. This additional armour appears, from examination of photographs, to be simply affixed to the outside of the angular turret armour originally visible. Though the plating is configured to look like a larger turret with bustle, there is no indication that any of this additional volume is used. At time of writing no explanation is known to the author as to why this was done, or why the vehicle was still in use at this time.

The modified T66 going through heavy mud. Note that the bustle appears to be a simple sheet plate attachment, not used for storage. Possibly a counterweight?

Above: ¾ view of T66 in a left hand turn. Note the steering of the centre wheels, and the gap between the round turret and the plates of the straight bustle.
Below: T66 fails to negotiate what the Ordnance Branch's caption terms "a medium obstacle."

3-Inch Gun Motor Carriages

The first wheeled 3-inch gun motor carriage saw work started in September 1916, consisting of a 3-Inch AA Gun M1917 mounted on a special four-wheel truck chassis, which was tested in 1917. In 1921 a convertible type 3-inch gun motor carriage, which could operate on either tracks or wheels, was tested in Aberdeen Proving Ground. In 1931, a 3-inch AA gun was mounted on a six-wheel truck chassis.

Because of the meager funds available for development, no other 3-inch gun motor carriage project was authorized until 1940, when the success of the German assaults forced a reappraisal of all fighting methods. At this time, interest in gun motor carriages was revived and to include calling for the use of a 3-inch gun. Although tracked vehicles would be the most logical carriage for a gun of this size, some wheeled programs were initiated.

3-Inch Gun Motor Carriage T15

In July 1941, the possibility of mounting a 3-inch gun on a Special Ford 4x4 chassis was considered. The following month, however, the layout was revised to provide for a 3-inch gun mounted on a special Ford 6x6 chassis using a Hercules engine instead of the original 4x4 Zephyr engine. The Ordnance Committee was planning to take action on the project, using the redesigned layout, but after further consideration recommended that the project be dropped. It was officially cancelled by the Ordnance Committee on 30th October 1941, before any military characteristics had been stated or a pilot built.

3-Inch Gun Motor Carriage T7

The history of the 3-Inch Gun Motor Carriage T7 is closely related to that of the Armored Car T13, on which it was to have been based.

Procurement of two Armored Cars T13, a modification of a commercially-designed vehicle known as a "trackless tank", was recommended by the Ordnance Committee in April 1941, in accordance with recommendations by the Armored Force Board. In giving its approval, the Adjutant General's office increased the quantity authorized to 17, and on 17 June 1941, further directed that four of these 17 vehicles be developed as 3-inch gun motor carriages, conforming as closely as possible to the military characteristics approved for the 3-inch Gun Motor Carriage T1 (M5) . This proposed vehicle was designated 3-Inch Gun Motor Carriage T7.

T13, turret reversed

In October 1941 it was decided that two of the four T7s authorized should be built as 105mm. Howitzer Motor Carriages T39.

After discussion of layout drawings of the 3-Inch Gun Motor Carriage T7, it was agreed to change the military characteristics. The revised characteristics, approved by the Ordnance Committee on 15th January 1942, increased the length, width, and height. The gun, recoil mechanism, gun mount, and gun shield were to be identical with those authorized for the 3-Inch Gun Motor Carriage M5.

Colonel Ray Montgomery of the Tank Destroyer branch visited the Reo Motor Car Company of Pontiac, Michigan, which was manufacturing the T13 during early May 1942. He described his thoughts in a letter to the Board thusly:

"It is difficult to dependably size up this Trackless Tank situation. Everyone connected with it seems to be an enthusiast, working hard to get the thing going. It is radical in design and is likely to be either a phenomenal success or a complete bust when it has had a fair and exhaustive test, which I think it is now going to get."

He concluded his letter with saying that "if a 3-inch gun for the T7 is not on its way to Lansing by about June 1st, I [will] take the matter up personally with Col. Christmas to obtain one from any source, by arbitrary action to divert from other use if necessary, and have it shipped by special truck to Lansing to avoid further delay in developing of the 3" gun phase of this Trackless Tank idea. This will be especially important if the acceptance tests turn out favourably."

Construction of the pilot models of the 3-inch Gun Motor Carriage T7 and the 105mm Howitzer Motor Carriage T39 was held up pending the completion and testing of two pilots of the armored car T13. These tests were conducted at Fort Knox in the summer of 1942 and showed that the chassis of the vehicle was unsatisfactory, whereupon work on the entire project, including the T7 and the T39, was suspended. The pilot armored Cars T13 were shipped back to the manufacturer, to be reworked at his own expense, the government agreeing to test them once more when the remodeling was completed. In September 1942, Headquarters, Services of Supply, directed that the project for Armored Car T13 be cancelled.

Acting on this directive, together with an unfavorable report of the Special Armored Vehicle Board and Memoranda from the Assistant Chief of Staff, G-4, and Headquarters, Services of Supply, the Ordnance Committee, on 24 January 1943, formally closed the projects for the 3-Inch Gun Motor Carriage T7 as well as for Armored Cars T13 and T13E1 and the 105mm Howitzer Motor carriage T39.

T13, right side, turret reversed.

3-Inch Gun Motor Carriage T55 / T55E1

3" Gun Motor Carriage T55E1

As a result of favourable performance at the Desert Training Center, Indio, CA., of an 8x8 vehicle known as the "Cook Interceptor," (After the Cook brothers who designed it; Cook and Allied Machinery were known as "The Interceptor Co") it was decided to use its chassis as the basis for an experimental 3-inch gun motor carriage. The Ordnance Committee, in August 1942, recommended building two such pilots, designated 3-Inch Gun Motor Carriage T55. This was formally approved in September 1942 and the Los Angeles Ordnance District was ordered to negotiate a contract with the Allied Machinery Company for the production of the two vehicles.

Proposed military characteristics called for use of the 3-Inch Gun M6 on 3-Inch Gun Mount M4, which were originally designed for the 3-Inch Gun Motor Carriage M5. The gun was to have a range of elevation from -10° to +15° and a traverse of 20° right and 20° left, a cal. .50 machine gun was to be mounted for antiaircraft and ground fire and provision was to be made for carrying a cal. .45 submachine gun, a cal. .30 rifle with grenade launcher, and three cal. .30 carbines.

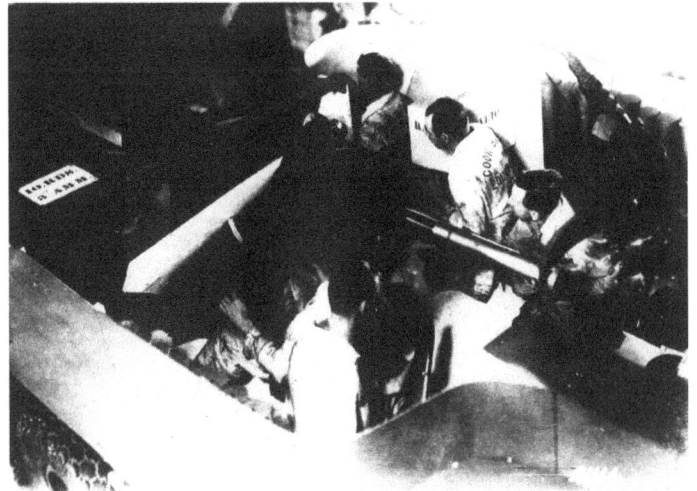

The T55 mockup. Note the position of the driver's shield, rotated behind him in the photo on the left. The photograph on the right shows the crew in position in the mockup

The vehicle was to carry 100 rounds of 3-inch ammunition as well as small arms ammunition. The frontal surfaces and side wings, including the gun shield and driver protection, were to be of ½ inch armour. With a crew of five men, the vehicle was expected to have a gross weight of 32,000 pounds and to be capable of a maximum speed of 50 miles per hour.

A distinctive feature of the proposed vehicle was its drive principle. The eight wheels were arranged in two bogies, each bogie powered by an individual engine. In the original chassis, one engine was located at the front and one at the rear. Design for the gun motor carriage, however, called for mounting dual engines at the rear, each with its own Hydra-Matic transmission. Initially, Lincoln Zephyr 125hp engines were to be used, later Cadillacs. The left engine supplied power or the front bogie and the right engine for the rear bogie. From the left engine, power was carried through its transmission to a transfer case, thence at a right angle through the center of a fifth wheel to the jack shaft differential, and then to the sprocket and out chains to the front and rear wheels. The vehicle was steered by turning the entire bogie about the fifth wheel by means of a hydraulic-actuated plunger. For close turning, steering was boosted by a hydraulic brake application to the front wheels, a valve permitting elimination of brake steering when not necessary, such as during high speed operation on a highway. From the right engine and transmission, power was carried through a transfer case to the sprocket and chains to the wheels. The jackshaft differential of the rear bogie was locked out of the system, so that there was no differential wheel action on the rear bogie.

The wooden mock-up was completed 5 October 1942, at which time it was decided that, in order to expedite completion of the pilot vehicle, the available Hydra-Matic transmissions would be used with a makeshift Spicer auxiliary transmission located between them and the right angle drive. The mockup showed fuel capacity of some 260 gallons (Two tanks of 50 and one of 30 in each rear sponson), giving a road range of some 800-1,000 miles. Tankage for 300 gallons of water was also envisioned, which seems a luxury use of space more likely to be used for ammunition or other stores.

The first pilot 3-Inch Gun Motor Carriage T55 was hastily completed and delivered to Aberdeen Proving Ground in November 1942 for testing by the Special Armored Vehicle Board.

Top: The original "Cook Interceptor" in testing in Indio, CA. Note the longer gap between the bogies compared to the T55 GMCs. Above: 3" GMC T55

Testing was quick. The Special Armored Vehicle Board submitted its preliminary report on 3rd December 1942.

In the opinion of this Board, a tank destroyer gun motor carriage must have cross-country mobility fully equal to that of a tank. Competetive tests show this vehicle to have inferior cross-country mobility to Gun Motor Carriage T-49. In addition the weight, length, width and height are all greater than those desired in a gun motor carriage for use as a tank destroyer.

None of the using arms represented on this Board (The Armored Force, The Tank Destroyer Center, and The Cavalry) desire Gun Motor Carriage T-55.

This Board finds no reason for further consideration of Gun Motor Carriage T-55 for service use or for further use or for further service test or to outline further developments to be pursued by the United States.

Above: T55 as it appeared in late 1942

The Board unanimously recommends termination of further development of Gun Motor Carriage T-55 for the United States Army

That would likely have have been the end of it, but there was a snag. The Army had contracted with Cook to build two vehicles, and was getting a second one whether they wanted it or not. Thus as a means of satisfactorily completing the contract for the manufacture of two vehicles, following instructions in OCM 20068 on 1 April 1943, the first pilot was returned to the manufacturer for certain changes, prior to placing it in the historical museum at Aberdeen Proving Ground.

A number of modifications were made to reduce the weight of the vehicle by some 11,000lbs and to obtain a more even weight distribution inasmuch as two-thirds of the total weight was carried by the rear bogie. The 3-Inch gun M6 was replaced by a 3-Inch gun M7 on a combination mount T49 modified to mount the M7 only. The engines were moved forward approximately three feet, with a resulting reduction in crew space. The crew compartment was rearranged to obtain a lower and more compact vehicle, the .50cal ring deleted, and the armour was reduced to ¼ inch.

The modified pilot, now designated 3-Inch gun Motor Carriage T55E1, was tested at the Ordnance Desert Proving Ground, Camp Seeley, Imperial CA., in comparison with other wheeled, half-track and full track vehicles.

Left: T55 again showing off the rotating system for the driver's shield.

This report was submitted 27 June 1943. On test in addition to T55E1 were 75mm GMC T66, Armored Car T19E1, being compared with Half-Track M3, light tank M5A1 and medium tank M4A3.

Above: T55E1 negotiating heavy mud. In the background is 75mm GMC T66 Right: Front view of T55E1 at Camp Seeley. Note that the driver has changed sides.

The Desert Proving ground saw both advantages and disadvantages in the drive principle of the T55E1. Driving strain was less, and the vehicle had excellent loading, under-vehicle clearance, and angles of approach and departure. When both bogies had traction and the engines were properly synchronized, the performance was excellent. However, when one bogie was out of traction only one engine was available to propel the vehicle, a handicap that was particularly noticeable on obstacles. Steering was unsatisfactory.

Because of these and other defects, the Desert Proving Ground concluded that T55E1 was not suitable for military service from either a design or performance standpoint. It pointed out that "the bogie suspension with chain drive as used thereon has merit and should be considered for multi-axle combat vehicles provided satisfactory steering could be obtained. "

Therefore, although the T55E1 was not used as a gun motor carriage, the experiments conducted with its chain drive principle proved very useful later, when it was used successfully on heavy trucks.

Next page: Three images of T55E1 under testing, and demonstrating the large range of motion of the bogies

At the conclusion of these tests, the vehicle was shipped to the Aberdeen Proving Ground for additional tests to obtain data of historical interest. It arrived in August with 3,009 miles on the clock.

These somewhat limited tests were completed by mid-January. The report dated 31st January 1944 primarily was a series of measurements of characteristics, but a run was made on the obstacle course where the long overhang and limited elevation proved to be a problem. In addition, the gun was proof fired, 8 rounds in total. The vehicle was considered stable enough, but there were a number of defects in the design.

Due to interferences in the cabin, maximum elevation varied depending on traverse: 17° at centre, 16° at right, and only 12° at maximum left. In order to get to maximum traverse, it was also required to move the driver and gunner's seats forward, the gunner would also have great difficulty in using the traversing and elevating hand wheels. The operating cam had been removed, making it a manual-loading gun only. The elevation grearing was loose enough that firing raised the gun 8°.

No recommendations were made by the report as the program had already been cancelled. The record shows the vehicle placed in the museum, but like many other vehicles supposedly preserved by Aberdeen for the historical record, it has since been lost.

Half-Tracks

75mm Gun Motor Carriage T12 (M3)

Early production 75mm GMC M3 with pedestal

At a conference at Aberdeen Proving Ground on 25 June 1941 between representatives of the Ordnance Department and the Assistant Chief of Staff, G-3, a project was initiated to mount a 75mm gun on a Half-Track Personnel Carrier M3, for testing as an expedient tank destroyer. Such a gun motor carriage, it was thought, would make the best possible use of old 75mm guns on hand, pending the production of a self-propelled weapon designed primarily for antitank use. It might serve an immediate function as a substitute weapon for a provisional tank destroyer unit during the forthcoming autumn maneuvers. It was proposed that the gun be mounted to fire forward and that it be served from the vehicle. Armour protection was to be the same as that of the vehicle, except that a shield was to be mounted on the gun if this could be done satisfactorily. The vehicle was to carry as much ammunition for the 75mm gun as possible. Each man of the proposed four-man crew was to carry his own rifle, 200 rounds of ammunition, and full field equipment. Provision was to be made for a radio receiver.

The gun selected was the 75mm Gun M1897A4, a modernization of the 75mm Gun M1897, often referred to as the "French 75" and regarded as the most effective light field gun used in World War I. Mounted on 75mm Gun Carriage M2A3, a two-wheel, split trail, pneumatic-tired gun carriage, it had elevations from -9° 14' to +49° 30' and could be traversed 30° left and 30° right.

Developing a muzzle velocity of 2,000 feet per second, APC ammunition fired from this gun had a maximum range of 13,870yds and could penetrate 2.5" of homogeneous armour plate at 1,000 yards. On 3 July 1941, the Adjutant General approved the project, and construction of the pilot model, designated 75mm Gun Motor Carriage T12, began at once. The design was developed at Aberdeen Proving Ground in the shortest possible time and without engineering studies to determine the ultimate application of the gun to a vehicle.

Military characteristics were amended to provide for a forward-folding frontal shield for protection of the driver and assistant driver, and a pedestal mount for a cal. .50 air-cooled machine gun. It was also directed that the 75mm gun should be equipped with the standard sight employed on the Gun Carriage M2A3. This sight was suitable for direct laying. The gunner was to perform all laying operations.

T12, part-way through conversion and missing the rear panels, rear fuel tanks, and floor ammunition containers

In the pilot vehicle, the gun was mounted on the frame of the half-track by means of a special base which replaced the wheels, trail, and other parts of the lower assembly of the gun carriage. The portions of the carriage from the trail axles up were retained.

A number of modifications were made in the basic Half-Track Personnel Carrier M3. To prevent interference when the gun was depressed, the glass windshield was removed, the armored shield for the windshield was made to fold down instead of up. Brackets were provided to secure the shield in its lowered position. A cut was also made in the top center of the armored shield to permit its being raised when the gun was locked in the travelling position. The rear vision mirror, formerly attached to a channel iron that held the top of the glass windshield, was moved to a channel iron at the left of the driver.

The side seats, gasoline tanks, sub-floor, and gun racks were removed, and substitute gasoline tanks were relocated on either side at the rear. A new sub-floor was added and also ten 4-round boxes for 75mm ammunition and one box for ammunition for the pedestal machine gun. Demonstration of the pilot model at Aberdeen on 16 July 1941 was so successful that authority was given for building 36 more such gun motor carriages for service tests. The pilot was modified as a guide for producing the additional vehicles.

Left: The vehicle in the same condition as the photo above.
Right: front view once the conversion was completed.

Nineteen individual ammunition containers with buffers and spring clips were arranged in the lower part of the gun mount. Rammer staff brackets were placed on the side wall to the right of the gun mount. The machine gun pedestal was moved from its previous position and bolted to the ammunition boxes. Repair chest brackets were mounted on the floor to the right of the gun. Two forward facing seats were installed, one on the front of each gas tank.

As tested at Aberdeen Proving Ground, the vehicle was operated by a crew of four men: a driver, an assistant driver, who was also the radio operator, a gunner, and a loader for the 75mm gun. One man of the gun crew also operated the machine gun mounted on the pedestal mount.

During travel the gunner and loader occupied the seats attached to the gasoline tanks. When firing, the gunner was at the left of the gun and the loader at the rear. Just before the gun went into action, the driver and assistant driver, who sat in the cab of the vehicle, dropped the windshield and crouched in their seats. The driver remained in his seat during firing of the gun in order to move the vehicle as required to facilitate traversing the gun beyond the traversing limits of the mount.

75mm GMC T12, left and front (shield down) views

The first report on the test of the pilot vehicle was dated 11 September 1941. By this point, limited production had already commenced, with a note that the pilot, and the first 36 production vehicles had no radios fitted due to lack of available sets.

The elevation and traverse of the gun had to be cut down due to interference between the gun and body of the vehicle. Traverse was cut down to 40 degrees, and elevation to approximately 30 degrees. The equilibrator rods interfered with ammunition stowage.

Slightly revised ammunition stowage with new spring clips

After initial measuring and photographing, the first rounds were fired on 19th July. The Proving Ground reported that the vehicle provided a very stable mount. When the gun was fired, the vehicle rolled back 3 or 4 inches, but returned to its original position with very little pitching motion. The main shock of firing the gun was taken up in the slight backward and forward roll, which was not considered objectionable. After 200 rounds, the vehicle was sent on an endurance run of 500 miles on the cross-country course at the aviation field, towing a 105mm howitzer and the vehicle weighted down to simulate a full load. Later in the testing program, another 400 or so rounds were fired at excess pressure to further stress the vehicle and gun. The testing was concluded on 15 August.

Still, despite the overall satisfactory result, Aberdeen did find some room for improvement.

The vehicle as manufactured had the standard gun carriage M2A3 shield. On future models it would be adviseable to enlarge the shield to give added protection to the crew of the gun, as during firing of the gun the driver's shield must be folded down. The shield should also be thicker to adequately protect the gun crew.

The original directive calls for a crew of four members. This crew should be increased to five members, to include, a driver, an assistant driver who is also the radio operator, a gunner, an assistant gunner, and a loader for the 75mm gun. This would give a flexible crew. There is provision in the vehicle for only two men in the back of the vehicle. It is suggested that another seat be added or that the additional man be seated on top of the 75mm repair chest.

Rear of the completed T12

Under the present directive, no mention is made of a machine gun being mounted on the pedestal in the rear. This vehicle needs a gun on this pedestal for anti-aircraft and for protection for ground personnel.

During proof firing of the gun, considerable trouble was encountered in the breaking of the headlight lenses when the gun was fired in the depressed position. A different location for the headlight would be more satisfactory.

This mount is considered to be a very good 75mm gun motor carriage combining, as far as present research on the subject has gone, the best in choice of vehicle and gun.

Pilot 75mm GMC T12, top view.

As a result of the test of the first 36 production vehicles at Aberdeen Proving Ground and by the 93rd Anti-tank Battalion, Fort Meade, Md, the military characteristics were revised somewhat, and the vehicle was standardized as the 75mm Gun Motor Carriage M3 on 21 November 1941, just two weeks before Pearl Harbor. A few of these vehicles had been rushed to the Philippine Islands and were used in the defence against the Japanese.

Military characteristics of the standardized vehicle called for the use of armour at the front, sides, and rear, as on Half-Track Personnel Carrier M3, and for a gun shield to give protection against cal. .30 armour-piercing ammunition at 250 yards and overhead protection against frontal attack by low-flying aircraft.

The pedestal mount of the pilot model, and ammunition containers displayed open

Several different gun shields were designed at Aberdeen Proving Ground. The first of these extended from the front of the body of the vehicle to a point about two-thirds of the way back, and was high enough to permit the gunner and loader to stand at full height. The high silhouette of this shield was considered objectionable, and a new shield, of approximately half the height and extending only half the length of the vehicle body, was designed.

The gun shield, as used on the standardized vehicle, was designed to traverse with the gun. The front and top plates were sloped to lessen the possibility of bullet penetration. This shield extended approximately a third the length of the vehicle body. Because of its low silhouette it was necessary for the gunner and loader to sit down in order to be fully shielded by it.

Left: The first two attempts at replacing the towed gun's shield were large fixed affairs The upper shield was photographed on 21st October, the lower shield on 12th November

While this was going on, Armored Force had acquired a few vehicles for use in the Louisiana and Carolina maneuvers of autumn 1941, the report being submitted on 22 December. It stated that the 75mm Gun Motor Carriage T12 (M3) was not completely satisfactory as a self-propelled weapon for armored field artillery, and should be used only as a guide in making a more satisfactory mount. The Armored Force Board agreed, however, that it could be used until a more suitable mount was constructed.

It is perhaps instructive to note the Armored Force's focus:
The test consisted in determining:
(1) Time necessary to enter and go out of action. (2) Time to execute 1600 and 3200 mil deflection shifts. (3) Cross country mobility of the mount. (4) Placing mounts in concealed positions. (5) Advantages over towed mounts, if any, in selection and occupation of position. (6) Any other pertinent information worthy of note.

Two sets of tests were conducted, one set using a battery of four vehicles during the Louisiana and Carolina Maneuvers, the battery being manned by the 27th Field Artillery Battalion (Armored), and at Fort Knox using inexperienced troops of the new 67th Field Artillery Regiment of 5th Armored Division. All the tests were related to the use of the gun in the traditional field artillery role, not 'tank destroyer' roles.

Above: Field Artillery Crew in Fort Knox

The Board commented: "The 75-mm Gun Motor Carriage, T12, although not completely satisfactory as a self-propelled Field Artillery weapon, has served to demonstrate some of the advantages and disadvantages of a self-propelled mount. The principal advantages demonstrated from this test are that the self-propelled mounts can occupy firing position more rapidly than towed guns and that self-propelled mounts can leave a firing position more rapidly than a towed gun. Self-propelled mounts and towed guns possess the same

75mm GMC T12 in Fort Knox, showing maximum elevation. The gun appears to be in full recoil position, though no explanation accompanied the photograph in the archives

degree of rapidity in initial laying and control in the firing battery. For direct laying, self-propelled mounts can be placed in firing position almost instantaneously. For indirect fire, they can be placed in firing position in less than one-fourth of the time required for towed guns."

The Armored Force Board reported that the number of seats was inadequate for a proper field artillery crew and that two more seats should be added. It recommended that two sheet metal stowage boxes be placed on the outside rear of the vehicle.

It proposed elimination of duplication in gun and vehicular accessories, such as hammers and wrenches. It reported that the vehicle was overloaded and that performance and durability were therefore diminished. It also found that the traverse provided in the mount was inadequate (Taking 2 minutes 18 seconds to conduct a 1600mil shift, vs 22 seconds for four towed guns).

The board also concluded that there was still a place for a 75mm assault howitzer in Armored Force units, as it had recommended in a letter 23 October 1941. But there was no mention of the use of the weapon in the anti-tank role, presumably it considered the tanks of armored units to be equal to the task.

In any case, in accordance with a directive from Headquarters, Services of Supply, the Ordnance Committee in April 1942 authorized modification of the vehicles then in production. The gunner's seat, pedestal mount, and cal. .50 machine gun, together with the provision for stowing the cal. .50 ammunition were removed. The telescope sight was raised approximately 2 inches and the slot in the shield modified accordingly. The gun shield vibration dampers were removed.

Above: 29th November 1941 (after the vehicle was standardised) the final version of the gun shield is shown

Because the standard "on-carriage" fire control equipment as originally proposed was unavailable, the Ordnance Committee in April 1942 authorized the use of substitute equipment. This included Telescope M33, Telescope mount M36, and Instrument Light M17. Produced hurriedly and in considerable quantity, 75-mm Gun Motor Carriage M3 was supplied to the British under the Lend-Lease program.

Calls were received for increased production, exceeding the available supply of 75mm Gun Carriages M2A3 used in assembling the gun mount. To produce a mount with equivalent characteristics, it was necessary to make use of 75mm Gun Carriage M2A2, supplies of which were available.

The Ordnance Committee in July 1942 designated the gun mount using the M2A2 carriage, 75mm Gun Mount M5, and the vehicle with this mount, 75mm Gun Motor Carriage M3A1.

A pilot model was delivered at Aberdeen Proving Ground where a ten-round proof-firing test was conducted, as well as a cross-country operation test of 200 miles in what the report noted was exceedingly rough conditions due to bad weather. Aberdeen concluded that the 75-mm Gun Motor Carriage M3A1 was as satisfactory as the M3 except that the maximum depression was limited to less than 7° as compared with the maximum depression of nearly 10° on 75-mm Gun Motor Carriage M3.

Comparison of the two gun mounts revealed little apparent difference. The M5 was placed 3¾" farther back on the base than was the M3 mount, a change that permitted the traveling lock to engage the traveling lock bracket when in the locked position.

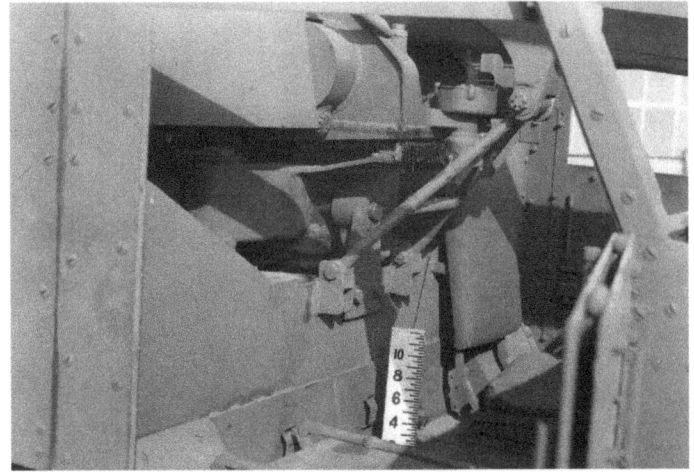

Comparison photographs between the mounts on the M3 (Top) and M3A1 (Above)

Above: A 75mm Gun Motor Carriage M3 and a 75mm Howitzer Motor Carriage T30
showing off their cold-weather windshields in tests of winterization equipment in 1943

A further series of tests was conducted with the M3, for the implementation of a computing gunsight. Two sights were evaluated, the T18 and T62.

Computing sight T18 mounted on the M3, and a closer view of the azimuth tracker side (right) and elevation tracker side (Below)

Designed to engage moving targets at ranges from 200 to 2,500 yards, operation of T18 required two crewmen, one for azimuth and one for elevation. The sight would compute lead angles in the horizontal plane so that the bursts would occur on the line of sight of the future position of the target, provided the range data was fairly accurate and provided the target continued to move to the predicted position during the time of flight of the projectile. The required lead was put into the gun during the four-second period between the time the starting lever was pulled and the firing flag in the sight fell. The sight would then come back to zero lead after each shot. The computer was in a case approximately 17"x10¾"x7", and was mounted on the sub-calibre mounting bracket on the gun.

The system proved reasonably accurate, with variations on range and speeds up to 20mph not affecting the accuracy of the system, at least once a 10% increase in the time of flight had been made in the computer. (It was theorised that there was some internal slippage). The catch was that four seconds of tracking seemed a little long, and a reduction to two seconds would have been preferred. In addition, the reset-to-zero seemed a little sluggish as well.

Left: Top view of the innards of Computing Sight T18

The alternate option, T62, was a one-man affair. Powered, like the T18, by a spring motor, it was designed to calculate lead at ranges from 500-2,500 yards. When the telescope was on the target, the computing mechanism was started by pulling the operating lever. After a few seconds (the exact time depended on the range to the target) the telescope jumped off the target. When the sight was again brought on the target, the lead would be put lnto the gun and thus be ready to fire. The lead remained in the sight until the operating lever was pulled again. The 18"x11"x21" sight was bolted to the left side of the gun.

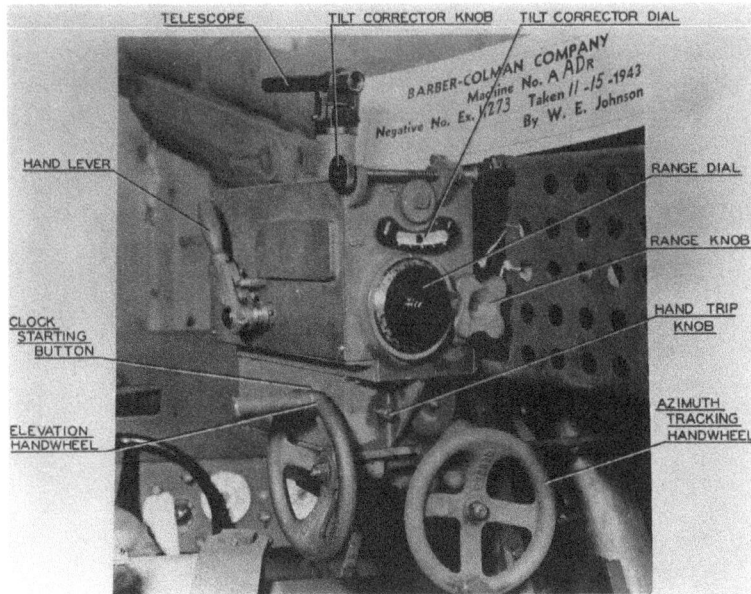

Components of the Computing Sight T62

T62 had its own deficiencies. Firstly, the gunner was considered to be overworked on anything other than level ground, the two-man system of T18 was preferred. The lack of cushioning for the head resulted in bruising; after 15 rounds, gunners started to flinch. Further, if the gunner decided that the lead in the computer was wrong, maybe due to his false input, there was no way to reset the lead calculation: He needed to wait until the clock ran down before the lead set to zero and he could re-start the tracking process.

The lack of a 'zero' range setting on both devices led to difficulty in boresighting. On the plus side, both systems stood up to the shock of firing.

In the end, it was concluded that neither sight was satisfactory for military use or to progress to service board tests. Such a system would be revisited at a future time, on a later vehicle: M18.

Above: An M3 on exercise in the US

In March 1944 upon standardization of 76mm Gun Motor Carriage M18, 75mm Gun Motor Carriages M3 and M3A1 were reclassified as Limited Standard. As quickly as troops could be supplied with the newer weapons, the M3 and M3A1 were withdrawn from service. In August 1944 the Ordnance Committee recommended that the vehicles be declared obsolete, that the artillery items be removed, and the chassis be converted to Half Track Personnel Carriers M3A1.

75mm Gun Motor Carriage T73

In early 1943 it was reported that the existing stock of 75mm Gun M1897A4, used on the 75mm Gun Motor Carriage M3 had approached exhaustion. Because of the success of the gun motor carriage, it was considered desirable to continue manufacturing it, using 75mm Gun M3, the standard gun for Medium Tank M4 series, which was then in production.

Authority was given in March 1943 for designing and manufacturing an experimental sleigh and making the necessary modifications to 75mm Gun M3 for mounting that gun on 75mm Gun Recoil Mechanism M2. It was proposed to mount the gun and recoil mechanism on a half-track vehicle in the same manner as on 75mm Gun Motor Carriage M3. Designations assigned were 75mm Gun Motor Carriage T73, 75mm Gun Mount T17, and 75mm Gun T15.

When the necessary modifications had been made, the vehicle was assembled and prepared for test firing. In the meantime, however, word was received that production of these mounts would not be undertaken, and so the tests conducted at Aberdeen Proving Ground were of a preliminary nature only.

75mm gun T15 on mount T17

The Proving Ground reported that the gun, recoil mechanism, and mount were generally satisfactory in the gun motor carriage. An interference was discovered, however, between the breech of the gun and the floor plate of the vehicle when the gun was 45 inches out of battery at an elevation of 26 or greater and at center traverse. An interference was also discovered between the breech of the gun and the gasoline tanks when the gun was 45 inches out of battery at maximum right or left traverse and at elevations greater than 3°25'. Aberdeen reported that the handle on the lanyard was jerked from the grasp of the gunner as the gun recoiled, and might possibly injure the gunner's wrist. It recommended that the gun, recoil mechanism, and mount be considered satisfactory after the difficulties had been corrected. Because of the development of greatly improved gun motor carriages, the project for 75mm Gun Motor Carriage T73 was discontinued.

57mm

The 75mm was considered something of an expedient AT gun, and the 57mm was considered to replace it as the standard anti-tank gun. As a result, we go out of sequence a little here, and go from the 75mm to the 57mm instead of the more usual progression of size elsewhere in this book.

57mm Gun Motor Carriage T48

The idea of mating the 57mm M1 with the half-track chassis was first given serious consideration in April 1942. The Ordnance Committee's Subcommittee on Automotive Equipment wrote:

Development of other motor carriages, based on the Half Track Personnel Carrier with guns of equal or greater power, has demonstrated the suitability of this vehicle for mounting the 57 mm gun. It is impracticable to obtain the necessary number of 57mm gun motor carriages specially designed for this weapon in the near future. It is therefore necessary to provide these guns mounted on a readily available and fully developed chassis.

Design studies have been prepared to mount the 57mm Gun, M1 on the Half Track Personnel Carrier, M3 in a manner similar to that employed in the 75mm Gun Motor Carriage, M3. A gun shield is provided in addition to the basic vehicle armor. If it is desired to utilize the maximum ground carriage traverse, provision may be made for doors in the vehicle side armor to clear the gun at full recoil.

The Ordnance Committee gave proposed characteristics of the vehicle and designated it the 57mm Gun Motor Carriage T48, giving approval for the construction of two pilots in OCM 18149 of 26th April 1942.

The 57mm gun M1 and pedestal was removed from the Gun Motor Carriage T44 for the first vehicle. However, it was decided that, although this pedestal in its then-current form was adequate for mounting the gun, it did not lend itself readily to production methods. In addition, the travel lock was not expected to prove satisfactory, even before tests.

Aberdeen proposed to the office of the Chief of Ordnance two competing drawings for a new, mass-production pedestal. One of them used a number of gun carriage components, thus making an interchangeable mount. The second option was a little more complicated, but would create a neater self-contained mount which didn't require the use of any gun carriage components. Ordnance decided to go with the former, arguing in favour of the simpler design and maximum use of 57mm Gun Carriage, M1 components. Since the quantity involved was expected to be relatively small (though in the end, almost a thousand vehicles were built) and and since the carriage parts would be available the first design was considered to be the most economical type. This new mount was designated 57mm mount T5.

The travel lock as finally installed

Consideration was first given to incorporating the travel lock in the windshield, but it was felt that on long runs it might be advantageous to drive with the windshield folded down, which would be impossible with such a travel lock. In its final design, the travel lock, which engaged the muzzle end of the gun, consisted of a clamp with a quick release mechanism and a pedestal base bolted to the top of the engine hood in front of the windshield. When the gun was in firing position the travel lock folded back. Removal of the travel lock from the body of the vehicle permitted the addition of a pivoted gunner's seat on the pedestal of the gun mount. This seat swung independently of the gun and the gunner could move with the gun merely by maintaining contact between his shoulders and the recoil guard on the mount.

The gun shield from the 57mm Gun Motor Carriage T44 was mounted temporarily and the vehicle was operated for 50 miles over a gravel road with a 3,650-pound load. Measurements indicated that no deformation of the frame was caused by this load. Thirteen rounds were then fired at normal and excess pressure at the extreme positions of elevation and traverse. No ill effects were noted on the gun or the vehicle.

The author has not seen any documentation indicating why the 57mm M1 from T44 was not, in the end, installed on the pilot vehicle as authorised, instead being replaced by a British 6-pounder Mk III, with its shorter, thicker barrel. Neither is there indication of more than one pilot actually being constructed. In any case, following these tests, a new shield was installed, and other improvements made.

The shield designed for this vehicle had ⅝-inch frontal armour and ¼-inch side and top armour of face-hardened plate. Because of the simplicity of the gun mount and the amount of space available in the vehicle, it was possible to use a low shield, the sides and top of which extended well back over the gun crew. In fact, the 57mm Gun Motor Carriage T48 with this shield had the lowest silhouette of any gun motor carriage based on the Half Track Personnel Carrier M3. The height from the ground to the top of the shield was only 90 inches.

The Ordnance Committee in drawing up the characteristics had requested a traverse of 35° right and 35° left, but interference between the gun and the side braces that supported the wingshields limited the traverse to 27 ½ to the right and 27 ½ left. The elevation was from - 5° to + 15°.

The same sighting equipment as provided with the 57mm Gun Carriage M1 was used. A sighting slit with an armored cover was included in the gun shield.

Because previous experience with guns mounted on Personnel Carrier M3 indicated that muzzle blast was likely to crack and break headlight lenses, the standard headlight brackets were replaced by brackets that permitted the headlight to be removed quickly when the gun was to be fired. Blackout lights which fit these brackets were also provided.

The armour-plate windshield, which folded down on the hood of the engine as in the 75mm Gun Motor Carriage M3, was placed on the vehicle for the protection of the driver when the gun was being fired. Because a space for the gun tube was cut between the two windshields, it was necessary to remove the top windshield braces and to substitute additional side braces. The windshield wipers were then mounted on the individual windshield sections.

Ammunition racks capable of holding 100 rounds of 57mm AP ammunition were placed in the rear of the vehicle. The racks were constructed in three sections, the top rack being entirely enclosed in a water-tight case and holding 20 rounds of ammunition which were ready for firing as soon as the front door was dropped open. The second rack, immediately below, was built in egg-crate fashion to hold 60 rounds of ammunition in fibre containers. These were held in place by retainers placed across the front of the racks. The remaining 20 rounds, also in fiber

Two shots of the interior of the fully stowed T48 pilot

containers, were carried in a floor compartment directly in front of the other racks.

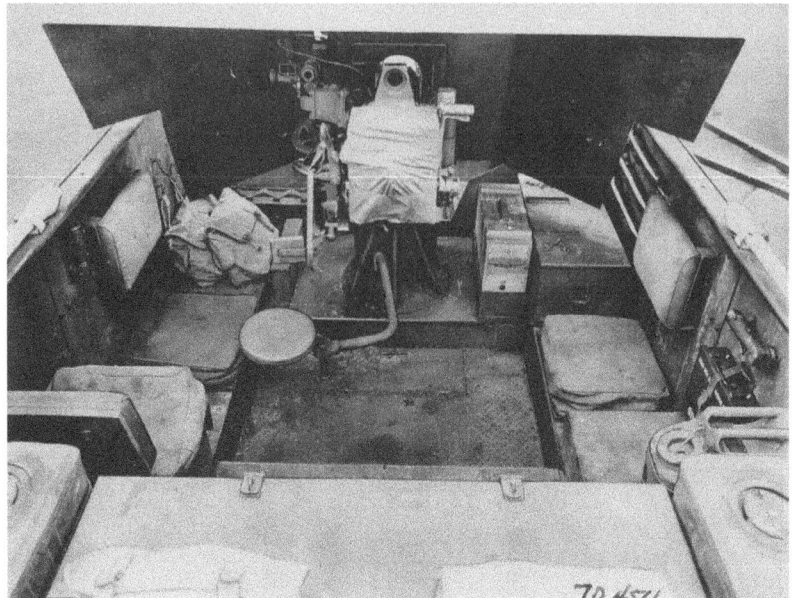

In July 1942 the vehicle was completed. An extended firing test of 100 rounds was conducted with the vehicle fully loaded. These rounds were fired at maximum positions of elevation and traverse and at service pressure. Measurements taken showed that no deformation of the frame occurred during firing. The only observed ill effect on the vehicle was a slight bowing-in of the center and two side hood plates due to the blast from the gun. This occurred primarily because the muzzle, when depressed, was very close to the top of the hood. Reinforcing angles were placed on all three plates to avoid this effect.

As a result of these tests, Aberdeen reported that the T48 was a satisfactory gun motor carriage and recommended that it be placed in production if there was a requirement for such a vehicle.

The conclusion from the testing was that most of the recommendations made in the report on 57mm Gun Motor Carriage T44 had been followed. The half-track vehicle used as a basis was capable of carrying the gun, ammunition, and other stowage, together with the five man crew, without overloading, and the firing platform was sufficiently large to allow the crew to service the gun at all positions.

It further concluded that the vehicle was superior to other gun motor carriages constructed from half-track vehicles in the following respects:

1. Lower silhouette - 90 inches above ground.
2. More protection from gun shield - ⅝" face-hardened armour plate for frontal protection.
3. Seat provided for gunner - pivots from gun base.
4. Removeable glass windshields provided for normal driving.

The tests were witnessed by members of the British Army staff interested in obtaining additional assault vehicles. Pleased by the results, they requested that manufacture of these vehicles be begun as soon as possible. Original plans had been to supply both US and British requirements, but subsequent changes in requirements resulted in their production for the British only.

The only change in the original characteristics requested by the British Army staff was that a British Wireless Set No19 be installed in the carriage instead of the Radio Set SCR-510 and that provision be made for carrying five British Cal. .303 rifles.

At the conclusion of the tests the 57mm Gun Motor Carriage T48 was sent from Aberdeen Proving Ground to the manufacturer and production was begun. After the British requirements were met (though produced for the British, the majority ended up sent to the USSR), all excess vehicles on hand were converted to half track personnel carriers M3A1 for US use. The T48 was declared obsolete in July 1945

Full-Tracked Vehicles

37mm Gun Motor CarriageT42

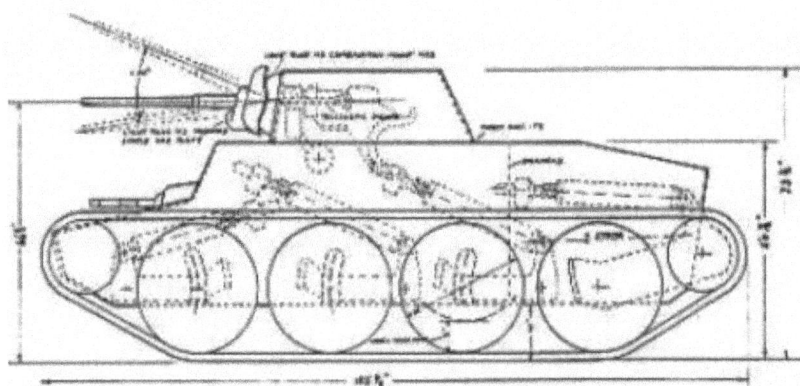

A letter from the Assistant Chief of Staff, G-3 (Operations), to the Assistant Chief of Staff, G-4 (Supply), 7 November 1941, recommended the T9 airborne light tank as a possible 37mm gun motor carriage for Tank Destroyer units. In order to improve combat efficiency, certain changes in design were suggested for incorporation in a pilot model. These included: increasing the hull ground clearance from 10 inches to approximately 14 inches; providing for individual springing and movement for each wheel; and lengthening of the hull and track base approximately 1 foot, to provide more space and increase stability.

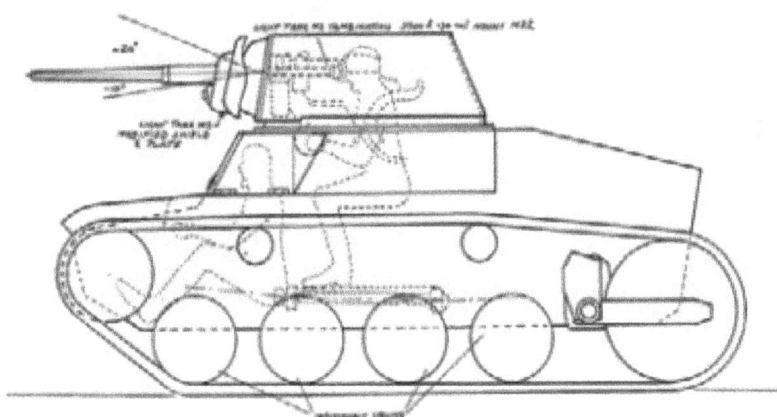

The letter recommended that the Chief of Ordnance contract for an additional pilot model of the Light Tank T9, modified, for test as a 37mm gun motor carriage for antitank use, and that this project be given a high priority to expedite construction.

A second letter from G-3 to G-4 in December 1941 submitted proposed military characteristics based generally on those being incorporated in the 37mm Gun Motor Carriages T22 and T23, with the additional requirement that the T42 vehicle be of the tracklaying type, using an independent

Above: Initial design of T42 Nov 1941. Re-designed Oct 42 to the large individual wheels shown in top drawing

suspension system of the Christie type. The proposed characteristics called for a 37mm Gun M6, a coaxial cal. .30 machine gun, and face-hardened armour, ⅞ inch thick on turret and front, ⅜ inch on sides, and ⅜ inch over engine compartment. The vehicle was to be approximately 7 feet wide, 14½ feet long, and as low as possible; the weight was not to exceed 6½ tons.

Layout drawings were prepared by the General Motors Corporation and in March 1942 a contract was placed with the Buick Division of General Motors for the manufacture of two soft-plate pilots.
At an Ordnance conference in April, objection was raised to the development of the 37mm Gun Motor Carriage T42, on the grounds that the 37mm gun was too light for field use. Subsequent discussion resulted in a decision to redesign the vehicle to permit mounting a 57mm gun.

The project was reassigned and the vehicle was redesignated as the 57mm Gun Motor Carriage T49. The T42 was never built.

57mm

57-mm Gun Motor Carriage T49

57mm Gun Motor Carriage T49, before final completion

After layout drawings for the 37mm Gun Motor Carriage T42 had been prepared, following a request of 31 March 1942, which led to a discussion between Generals Moore and Barnes, the Ordnance Committee, in April 1942, recommended use of the more powerful 57mm gun with the T42 chassis. Two pilot models of the proposed vehicle, designated 57mm Gun Motor Carriage T49, were authorized. In these a 57mm Gun M1 was to be mounted coaxially with a cal. .30 machine gun in a 360° power-traversed turret, and an additional cal. .30 machine gun was to be carried in a bow mount.

The mock-up at the Buick plant in Michigan

Maximum armour thickness was to be ⅞ inches, with the turret walls sloped to provide greater protection against enemy fire. With gross weight estimated at about 24,000 pounds, a maximum speed of 50 m.p.h. was anticipated and a ground pressure of 8¾lb./sq.in. Two Buick "60" engines were to power the vehicle. Two GM 6-cylinder truck engines were briefly considered for the second pilot.

A represenative of Armored Force visited the Buick plant in Flint, and his letter of 24th April, after viewing the wooden mockup under construction indicates some some of the unresolved questions:

As designed, T49 was to have dual driver's controls, but if the radio was to be mounted in the front right hull, it would interfere with the A-driver's ability to drive. Did Tank Destroyer Center have any policy as regards dual control?

It was suggested that for TD use the crew be reduced to four: Driver and commander in front, and gunner and loader in the turret: With the commander in front with the radio, a smaller turret could be used. In passing, he also mentions that the first T49 would likely be designated a light tank, with the second pilot to be built as a tank destroyer.

Two photographs of the pilot under construction. The officer is believed to be CPT Cushman

General Bruce of Tank Destroyer command, however, had his own priorities, and lobbied hard for the first prototype to be delivered to his command at Temple, TX. A letter of his dated 20th May already indicates that they had started looking at mounting a 3" gun into the chassis. The other point to note was that the services were very concerned about a rubber shortage, another reason for preferring a tracked vehicle to a wheeled one. However, he did indicate that he was keeping an eye on the T18 armoured car as a possible 3" mount.

A letter sent by him to General Christmas on the 25th is probably worthy of placing here. Extract follows:

I was disappointed over missing you in Detroit the other day. I have been wanting to talk with you personally for a long time but just haven't found the opportunity. There have been several items which we have indicated in official communications that somehow or other seem not to have reached everyone concerned. For example, the possibilities of the light armored car T22E2 as a several purpose vehicle is not known by many people.

The 57-mm gun motor carriage T49 (Christie-GMC) appears to have excellent prospects for what we want in the self-propelled mount for the 57-mm gun. I don't believe it has been clearly brought to the attention of the designer that we would be willing to sacrifice the 360 traverse for either the 57-mm gun or the 3" gun in order to decrease weight, to increase speed, and to make a simpler design for getting into production. In other words, we do not have to have a true turret, and in fact do not want a closed turret. A gun that can traverse and fire to the front and to the rear, or a gun that can fire to the front or to the rear, will be satisfactory. I understand that if you do not have to fire to the flank you can cut down the width and thereby the weight, and certainly can cut down the turret weight. Another point in our characteristics is that we call for protection against caliber .50 in front. If we can gain speed by reducing weight of armor in front to the extent that it will afford protection against only caliber .30 guns I would be glad to sacrifice this.

I understand there has been some talk about changing the name of the 57mm gun motor carriage T49 to a tank. I fully realise we have got to win a war, and in fact whether we call the thing a "tahootie" or tank or anything else won't win the war. There is a little point, though, of primary interest and morale. I feel that our tank program is coming along in fine shape. Certainly the M5 is a good tank. The M4 medium contains advanced thought. And the tank which you have developed at Rock Island, which weighs about 20 tons, is an excellent product. In other words, the tank situation seems well in hand.

To date, except for the T22E2 which has been somewhat modified to make a reconnaissance vehicle suitable to several agencies in the Army, we have no strictly newly developed tank destroyer weapons.

I therefore would like to see the gun motor carriage name reserved for us, and that the Christie GMC development is primarily for tank destroyer purposes, both for mounting the 57mm and possibly the 3" gun. Later if this weapon proves to be a better tank than what we have, well and good. We cannot be 'dogs in the manger', and of course we will have to call it a tank and give it to the Armored Force. In the meantime, though I earnestly hope that we can concentrate on this mount for our own purposes.

I never did give you a specific answer on another question that came up at our last conference at Aberdeen, and that was whether or not I would be willing to have the crew on the ground when firing. My answer is yes. I prefer to have the crew in the vehicle with protection against caliber .30 machinegun fire, but again if it is a question of speed I am willing to sacrifice the desirability of carrying the crew provided they are somewhat protected from machinegun fire in action. They should be in rear, however, and not sticking out like sore thumbs as in Cletrac. [Ed- See 3" GMC T1/M5]

Artist's impression of T49, maybe to be called a "Tahootie". Note that it is fitted with two .30 cal machineguns, making it much more a light tank than Tank Destroyer.

The letter gives a good insight on both Tank Destroyer thinking, and the frustration that they were facing at a perceived lack of attention being given to their needs.

As a further example of the thinking, Col Montgomery, also of the TD board, visited Flint in early June. He emphasised that in its then-current design, the vehicle *was* a light tank. Notably, he also made a number of comparisons to the T4 Christie tank, viewing it as a development with a number of improvements, to include a spring-loaded idler to retain tension, new track with bushings, and return rollers. The turret and basket was to be identical to that of the medium tank T7, except of welded construction, not cast.

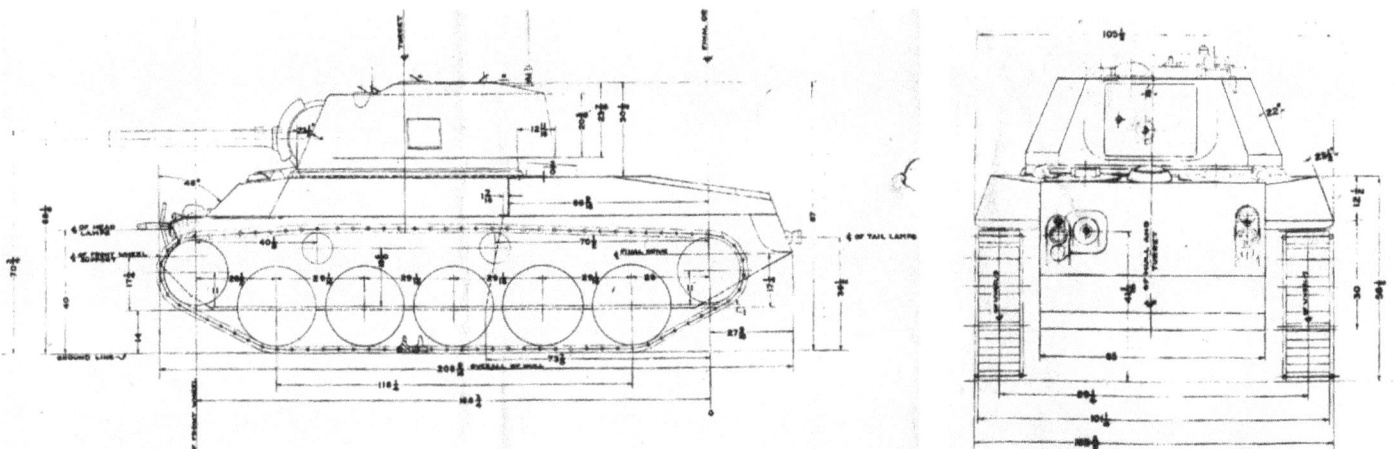

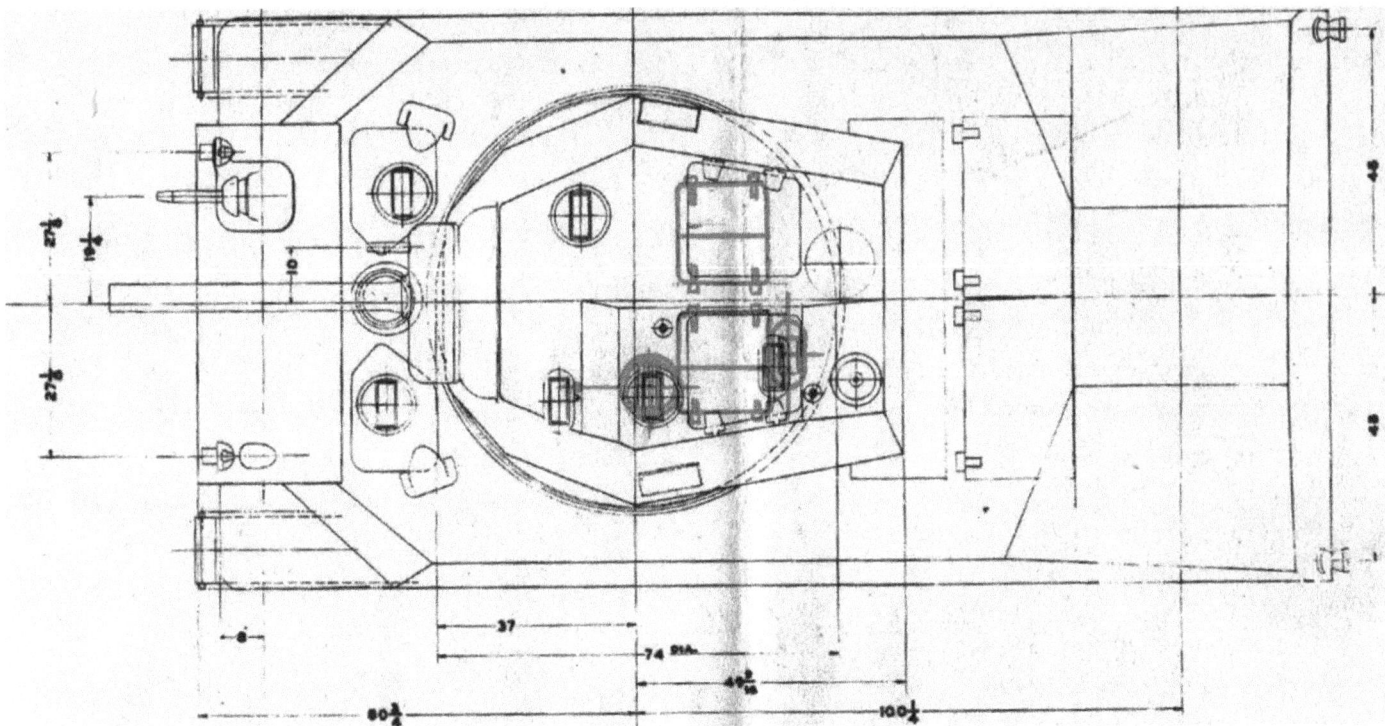

The Tank Destroyer Board were not happy. The President, COL Ross, wrote to Gen Bruce stating that the 57mm GMC T49 was not suitable for tank destroyer use even before the vehicle had been completed. His memorandum indicated the following changes should be required:

1. A change in the driving position to one similar to the M8 armored car, instead of use of periscope.
2. An open roof on the turret. Both these two were for visibility. The gun commander was to be in the assistant driver's position.
3. Reduction to a four-man crew, two in the turret and two in the hull. As the vehicle was to be issued to replace M6 GMCs, which had 4-man crews, there would thus be no manning problems.
4. Deletion of bow gun. The gun commander should be too busy fighting tanks with his vehicle than to be personally shooting infantry with a machinegun.
5. Move radio to the hull, to eliminate demonstrated problems with slip rings.
6. Installation of an intercom system. It might get noisy in a full-track vehicle.

That was on 8th June. A week later, the Tank Destroyer Board received a letter from its liason officer at Armored Force in Fort Knox, CPT Cushing. He, too, believed that T49 required changes to be suitable for tank destroyer use.

The proposal was more drastic: That the turret be simply removed, and the 57mm mounted on a pedestal, probably using the mount from the half-track T48. By confining the field of fire to the forward 90 degrees, the vehicle's hull could likely be reshaped for better ballistic protection combined with a decrease in weight. The estimate was that in addition to losing the turret (2,600lbs), the overall weight of the vehicle could also be reduced by elimination of the turret race (690lbs), basket (285lbs), and traversing mechanism (175lbs), to be offset only by the weight of the 57mm pedestal and gunshield. Other than this proposal being filed away in the archives, there is no indication of any effort being undertaken to investigate it.

Stowage was expected to be a problem. Due to the shallow dimensions of the sponsons, 'normal' racks which required lifting ammunition up and then out would not work. Buick devised as a result a system of five-round belts in drawers. Two belts could be laid one above the other in each drawer, and when a round was removed as it was pulled out, the belt clipped so that the remaining rounds would not rattle around.

GMC T49 under test.

No changes were made. Even just opening the roof of the turret was beyond the authority of TD branch; the change would have to come from Washington. In order to not further delay the project, it continued as designed. The first pilot model, except the turret gun mount and gun, was completed by the Buick Motor Co., Flint, Mich., in July 1942 and was put through preliminary testing while awaiting arrival of the gun and gun mount from Rock Island Arsenal. When these had been installed additional testing was conducted at General Motors Proving Ground. Automotive tests indicated a loss of power sufficient to reduce the anticipated maximum speed of 55 to 60 m.p.h. to only 53 m.p.h, much slower in turns. When it was found that the torque converter was wasting too much power, it was planned to substitute two hydra-matic transmissions.

The suspension system, including rear drive, showed great promise. It was reported that the vehicle demonstrated superior riding characteristics and provided an improved gun platform. The suspension apparently reduced track throwing and permitted the operation of the tank at greater speeds over rough ground. A ¾ mile race was held against an M5 light: Though off to a slow start, the T49 had caught up by the first corner, took the inside, pulled away a few lengths and was several hundred yards ahead when it crossed the line.

GMC T49 takes a sideslope. Compare with the first photo in this entry, and one can see the persicope guards have now been installed onto the turret

Already, however, the testing reports were being written with caveats about expected performance with the heavier 75mm gun and turret when installed.

In the background, the Tank Destroyer Command had been studying the respective characteristics of the 75mm and 57mm ammunition. A letter from General Bruce to Army Ground Forces on 2nd July, with the first pilot barely delivered that week, stated:

Information just made available to this headquarters regarding comparative characteristics of the 75mm and 57mm ammunition, indicates that the presently available ammunition of the 75mm gun is considerably superior to that available for the 57mm gun. It also appears from the information that the 57mm APC projectile, if and when developed, will still be inferior to the 75mm projectile at ranges in excess of 500 yards. It is believed that most tank destroyer action will be at ranges of over 500 yards. Decisions made by this headquarters regarding standardization of the 57mm for tank destroyers were based upon the assumption that APC ammunition was being developed. Ballistic tables received from various sources and on file at this headquarters, give the armor piercing characteristics for 57mm APC projectile, and from this it was assumed that it would be available.

In view of the information now available, it is believed that there is nothing to be gained by continuing development of the 57mm gun as a tank destroyer weapon. Its only advantage over the 75mm gun is in the time of flight of the projectile, and that advantage is more than offset by lack of armor piercing characteristics at ranges over 500 yards, size of the round, and complication of supply. The available 75mm guns may be used pending development of suitable mounts for the 3" gun and larger guns.

As a result, and given that indications were that the 75mm gun M2A3 could be mounted on T49 without any significant changes, Gen Bruce recommended a change in the required characteristics to include a 75mm gun, and also that the designation of the mount be changed.

In accordance with the desires of the Tank Destroyer Command, the Ordnance Department directed that the second pilot should be developed to employ the 75mm gun M3 in an open-top turret. Ordnance Committee action, approved in December 1942, designated the modified vehicle 75mm Gun Motor Carriage T67. The first T49 pilot vehicle was returned to General Motors Proving Ground so that the 75mm gun could be installed. Both pilots were later sent to Aberdeen Proving Ground for testing and various improved features were incorporated in the vehicle later standardized as 76mm Gun Motor Carriage M18.

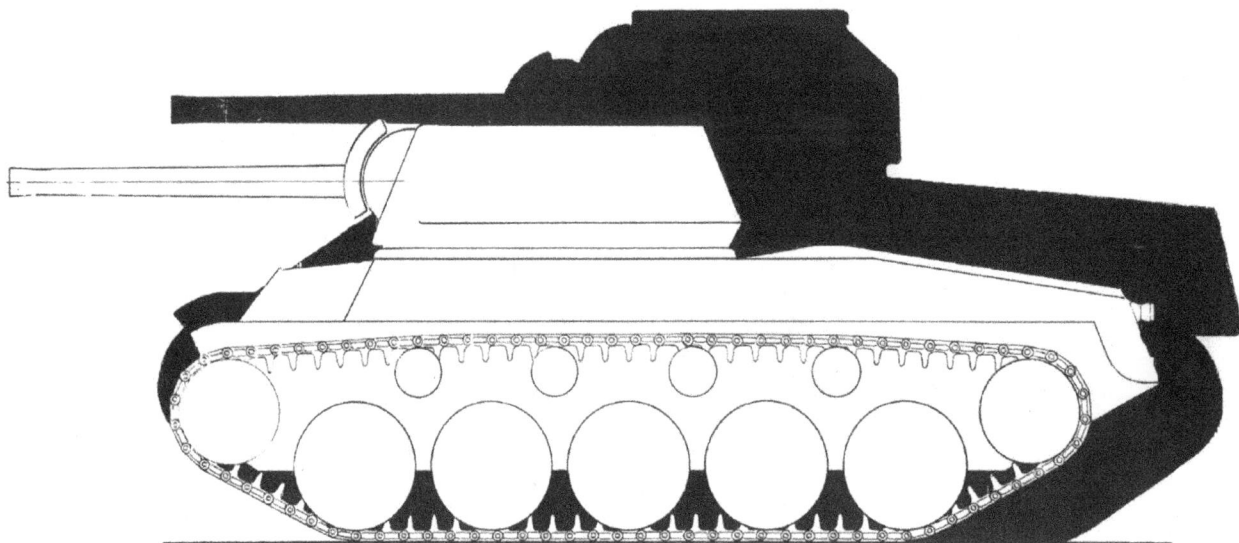

COMPARATIVE SILHOUETTE OF T-49 AND M-4

75mm

75 mm Gun Motor Carriage T29

This designation was assigned to a proposed 75mm gun motor carriage consisting of a modified turret of Medium Tank M4 mounted on the hull of Light Tank M3. According to layout drawings prepared in the latter part of 1941, the turret was to be thinned down and the rear section was to be removed.

Because no interest was shown by the using arms, the project was dropped at the request of Services of Supply without its ever having gone beyond the initial paper work.

75 mm Gun Motor Carriage T67

Left side and front views of T67, turret to the rear

After one pilot of T49 was built, the Tank Destroyer Command indicated that it preferred a 75mm gun on the vehicle. It was agreed, therefore, that the second pilot, then being manufactured, should be provided with a 75mm Gun M3 in an open-top turret. Ordnance Committee action in October 1942 gave military characteristics of the proposed vehicle, which was designated 75mm Gun Motor Carriage T67. The history of T49, T67 and T70 is entirely intertwined, with various references to include "76mm GMC T49" and "76mm GMC T67", as changes in the vehicle's characteristics occurred faster than Ordnance Committee action could keep up. The dividing points between T49, T67 and T70 are thus rather arbitrary.

The pilot model was delivered at Aberdeen Proving Ground in November 1942. It was powered by two Buick Series 60 engines, which operated through a torque converter and a manually controlled three-speed transmission. The turret and the 75mm gun mount were the same as those used on 75mm Gun Motor Carriage T66. Although it was intended that the T67 would have, if produced, thinner armour than T49, for simplicity of production and to get the testing started quickly, the turret was mounted directly onto to the T49 hull (without bow machinegun). In order to replicate the weight correctly, however, in testing the vehicle would only be loaded lightly during tests.

Because preliminary testing indicated that the vehicle lacked sufficient power, studies were made of a dual installation of the Continental R6572 engine, an in-line, liquid-cooled engine that was standard on High-Speed Tractor M5. This Continental engine had a piston displacement of 572 cu. in. as against the 320 cu. in. of the Buick engine. The substitution was not made, however.

The Special Armored Vehicle board, also known as the Palmer Board, in a preliminary report dated 3 December 1942, stated that given their opinion that a tank destroyer should be a gun motor carriage with mobility greater than that of a tank, ample space for the crew, not weigh over 20 tons, and have sufficient armour to stop small arms and fragmentation, T67 was the vehicle which

75mm GMC T67, using the independent coil spring suspension.

most closely fit their requirements. The board thus recommended that T67 development be continued, but that a heavy-duty gasoline engine currently in service with the Army be utilised, and the production be expedited to fill a critical need in the fighting equipment of the US Army.

These changes were incorporated in a new vehicle, which, in accordance with the desire of the Tank Destroyer Center, was equipped with a 76mm gun, and standardized as 76mm Gun Motor Carriage M18 (See subsequent entry for T70).

The official record indicates that after work had been started on the new vehicle, the second pilot of 75mm Gun Motor Carriage T67 was completed, and this was sent to the General Motors Proving Ground for testing of the track and suspension system. It goes on to say that results of these tests were so satisfactory that a individually sprung type of system with torsion bars substituting for coil springs was adopted for the 76mm gun motor carriage. However, this author has not personally encountered either photographs or test reports of this second vehicle, but records found in Tank Destroyer Board files indicate this occurred around late DEC42.

Photographs of T67 with the gun forward seem hard to come by. Indeed, photographs of T67 with the 75mm gun are also hard to come by; all photographs are of the vehicle after conversion to 76mm at the end of 1942. Note the sign still says 75mm though.

Having served as a phase in the development of an "ideal tank destroyer", and having provided a means of testing the components of the 76mm gun motor carriage, the project for the 75mm Gun Motor Carriage was formally closed by Ordnance Committee Action in January 1943.

75mm Gun M3 on 75mm Howitzer Motor Carriage M8 Chassis

On 30 January 1943 a 75mm Howitzer Motor Carriage M8 on which 75mm Gun M3 had been mounted in place of the 75mm howitzer was demonstrated before a group of War Department officials at Aberdeen Proving Ground. Although the vehicle had not been assembled primarily for use as a tank destroyer, the demonstration was so satisfactory that, at the request of a representative of the Tank Destroyer Command, it was shipped the next day to the Tank Destroyer Board, Camp Hood, Tx, for tests to determine its suitability for tank destroyer use. Under the informal designation, 75 Gun Motor Carriage M8 the vehicle was tested by the Tank Destroyer Board in February 1943; and then was transferred to the Armored Force Board, Fort Knox, Ky.

The vehicle had been modified to determine the feasibility of increasing the firepower of light tanks. To obtain the information promptly, the howitzer motor carriage was used. The vehicle had a light tank chassis and a semi-open turret in which the 75mm gun could be mounted more readily than if Light Tank M5 with closed turret were used.

In order to accommodate the gun, the front section of the turret was cut away, the opening being covered by the gun mantlet, and the cal. .50 machine gun ring was removed. Because the plate cover over the rear of the turret was also removed, a full open-top turret resulted. To permit the gun to be mounted on the recoil mechanism of the howitzer, an adapter was provided that made necessary the removal of the closing spring and housing on the gun and thus prevented the automatic closing of the breech block. The mounting was of an expedient nature only, but in spite of many admitted deficiencies, the modified vehicle performed excellently in preliminary tests at Aberdeen.

This page: Two views of the 75mm Gun M3 on 75mm HMC M8 Chassis

The complete set of conclusions was as follows:

The cross country mobility of the Gun Motor Carriage M8 was slightly superior to the 3" Gun Motor Carriage M10 and the 75mm Half track M3. On hard, rolling, cross-country terrain the 75mm Gun Motor Carriage M8 had no advantage over the 75mm Half-track M3, but both vehicles were superior to the 3" Gun Motor Carriage M10. On rough, uneven terrain, the 75mm Gun Motor Carriage M8 had no advantage over the 3" Gun Motor Carriage M10 yet both were superior to the 75mm Half-track M3.

Replacing the 75mm howitzer in the turret of the Gun Motor Carriage M8 with the 75mm gun had no major effect on stability and handling of the vehicle.

The M8 was definitely inferior to the M10 and M3 in the characteristic of crew comfort. On cross-country operation the M8 consistently threw the crew off balance, frequently injuring the men. Short, jerky movements of the vehicle were pronounced within the turret.

It was determined that the M8 could sustain a road rate of 30 m.p.h. over primary and secondary roads through hilly terrain. The rate was superior to that of the M10, and would enable the vehicle to remain in a column of wheeled trucks.

The braking characteristics were excellent. The vehicle could be stopped in 28 feet when traveling 20mph and 84 feet at 35mph.

The driver's vision from the driver's seat was limited in comparison with that of the M10 and M3. The dead space in front of the vehicle was measured at 21 feet.

The acceleration capacity of the M8 was superior: A speed of 30mph could be attained in 19seconds, travelling a distance of 585 feet. 20mph could be reached in 9 seconds at a distance of 186 feet.

The tactical efficiency of the M8 was equal to that of the present expedient mounts. The employment of the M8 in tactical problems showed no marked superiority over the M10 in negotiating rough terrain, and no superiority over the half-track in traversing hard, level ground. The small size and more rapid acceleration gave the M8 initial mobility advantage over the M10, but on continuous operation this advantage was neglible.

Comparative photo of "M8", M10 and M3 Gun Motor Carriages

The weight of the M8 at 15 tons gives it a pronounced tactical superiority over the M10. This would permit the use of most highway and military bridges that would be prohibitive to the 30½ tons of the M10.

The 4½ foot turret ring of the M8 rigidly confines the movement of the turret crew of three men. This feature interferes with the efficient use of the 75mm gun.

Shot of the primary ammunition stowage on "M8 GMC"

The vehicle was handicapped for going into action by the necessity of stowing the gun to the rear in the gun rest. This was mandatory, as the turret is provided with a single turret lock. This was considered too light to prevent breakage of the traversing mechanism if the gun should swing free during cross-country operation. Similarly the only way to avoid damage to the elevating gears during movement of any notable distance was to use the rear gun rest.

Stowage of 44 rounds was considered inadequate: Two hull sponsons were used, with 9 rounds in each compartment. Twenty additional rounds were placed on the floor of the turret compartment in an upright position, and six additional rounds were set in ready racks in the turret rear bulge. No space remained for personal equipment.

The vehicle was considered by the Tank Destroyer Board only as an expedient gun motor carriage that would be issued until such time as 76mm Gun Motor Carriage T70 (M18) could be placed in production.

The gunner's position in the "M8 GMC"

The report pointed out that the other two GMCs in service with the TD branch at that time were also expedient, so it would be a third one. It also pointed out that the M8 used 80 octane fuel, higher than that used of the other tank destroyer branch vehicles, excepting the M10s which used diesel. The Tank Destroyer Board was not enamoured of adding a third fuel type to their logistics mix, especially not for an expedient vehicle. Given the amount of effort required to fix the numerous problems in the M8 to bring it up to serviceable standard for field use, it was presumed that production of the expedient vehicle could not begin any earlier than that of the 76mm gun motor carriage T70 anyway, and as far as the Tank Destroyer Board were concerned, that was the end of that.

The Armored Force received the vehicle and tested it "To determine the suitability of and whether any requirements exist for this vehicle equipped with 75mm Tank Gun, M3, for Armored Force use." Although Armored Force never gave it a catchier name than the official "75mm M3 on HMC M8 Chassis," some enterprising folk at Aberdeen did stencil "M8A1" on the side of the vehicle. This was just as unofficial as TDB's "M8 GMC," so points to Armored Force for keeping things official. Armored Force seemed to be less enthralled with the vehicle than Aberdeen.

"The deficiencies as listed in Appendix 'E' of this report are so numerous and of such serious nature as to render the subject vehicle unsatisfactory for use in the Armored Force."

For those curious as to what deficiencies Appendix "E" listed, they were:

-*Introduction of this gun in the turret has reduced crew to four (4) men.*
-*Installation of this gun in M8 turret interferes with efficient service of the piece.*
-*The positioning of the fire control equipment with reference to the gunner is unsatisfactory in its general layout.*
-*With the gunner in firing position using 12" depression in the floor 45° right and left is the maximum traverse that can be obtained.*
-*The M3 tank gun used in conjunction with the 75mm M1 A1 Howitzer gun mount is not as accurate as when used with M34 or M34A1 gun mount in the tank.*
-*Turret is grossly out of balance.*
-*The gun mantlet interferes with the field of the direct telescope.*
-*Gun cannot be fired without taking hands off the controls.*
-*Firing solenoid does not function properly.*
-*Stowage facilities are generally unsatisfactory.*

75mm M3 on HMC M8 Chassis, with the external travel lock engaged. Due to the mount being more a proof of concept, developing a solid internal travel lock was not considered an efficient use of resources.

Mechanically, Armored Force had little issue with the vehicle, with mobility proving about the same as M8. Fuel consumption, for the curious, was 1.9mpg on road, 1.4mpg cross-country. The 54" ring still permitted no roof and no room for anti-aircraft protection, which was found objectionable. They also weren't so keen on the loss of the semi-automatic breech operation. Overall, though, they did like the idea of a light tank mounting a 75mm high velocity gun. Armored Force's view of such a weapon was that:

Front right view of the 75mm M3 on 75mm HMC M8 Chassis. (No wonder TD Branch invented a short GMC name for it.)

(1) It would greatly increase the fire power of our armored units.
(2) With improved fire control equipment, it would enable accurate fire to be delivered at ranges which are now prohibitive.
(3) It would provide an extremely mobile, high powered weapon of good accuracy, at long ranges.

Expedited design should be carried out with the idea of developing a satisfactory light tank w/75mm Tank Gun. The Light Tank, T-24, being developed by the Ordnance Department at Cadillac Division of General Motors shows excellent promise and should be expedited.

As a result, the final recommendation from Armored Force was "that no further consideration be given to the 75mm Howitzer Motor Carriage, M8, w/75mm Tank Gun, M3, and that the development of the Light Tank, T-24, be expedited."

It is perhaps a point of interest to compare the positions of the Armored Force with the Tank Destroyers when it came to the matter of crew configuration. The Armored Force, though it considered two-man turrets to be unacceptable, took the vehicle at face value: Manufacturer says '4', there were 4 seats, so they tested it with two men in the turret. Tank Destroyer Board, however, took the effort to try to figure out how to fit three men into the turret, manufacturer's specifications notwithstanding, and it came up with a couple of potential configurations, with some specifics given in the report:

1) The employment of three men in the turret of the M8 was tested by trying all possible arrangement of the turret crew.
2) It was found that positioning the gun commander on the left of the gun interfered with efficient operation of the loader.
3) The sight bracket of the telescopic sight was moved forward 4 inches, thus allowing the gunner to move forward to some extent. The right section of the floor in the hull compartment was removed to allow the gunner to assume a standing position. This lowered his feet 12 inches below the compartment floor. In such a position the gunner's ability to traverse the gun was limited to 90° to the left and 45° to the right.
4) The duties of the gunner and gun commander in serving the piece are relatively static as to position. Therefore, the commander was placed behind the gunner. This permitted full freedom of action to the loader.

98

5) The position of the gunner is cramped by the elevating and traversing handwheels. The position of these wheels is at the right shoulder of the gunner and places his arms in an awkward position. The axis of the sight is 6½" from the shoulder guard of the gun, thus placing the gunner's head uncomfortably close to the guard.
6) The results of the various arrangements of the three men within the turret emphasized the difficulties of three men serving the gun with any degree of efficiency or safety.

Above: Two of the crew configurations tried out by the Tank Destroyer Board. On left, the commander crowds the loader. In order to fit the commander in his 'traditional' place behind the gunner (right), modifications needed to be made in the turret.

Left: This photograph from the Armored Force report shows that the vehicle arrived in Fort Knox without being returned to the original configuration, the gunner remaining standing. The purpose of the seat so far behind him must have been confusing for Armored Force given their adherence to the concept of the vehicle having a four-man crew.

Because of the adverse report of the Tank Destroyer Board, and because it was also understood that the Armored Force had no requirement for such a vehicle, the Ordnance Committee recommended in June 1943 that the project be terminated. This action was approved 24 June 1943, and the vehicle was returned to Aberdeen Proving Ground to be reconverted to a howitzer motor carriage.

3-Inch Gun Motor Carriages

3-Inch Gun Motor Carriage T1 (M5)

A fully stowed 3" Gun Motor Carriage M5

In 1940, the Cleveland Tractor Company submitted a design for a gun motor carriage based on its high speed tractor MG2, which had been tested at Aberdeen Proving Ground with generally satisfactory results. This tractor, standardized in February 1941, became the standard air corps tractor under the designation 7-Ton High Speed Tractor M2. Although the original design had been prepared for a 75mm gun motor carriage, the Ordnance Committee considered the vehicle suitable for mounting a semi-automatic 3-inch gun with a muzzle velocity of 2,600 feet per second, which was then under development. In December 1940, the Ordnance Committee gave proposed military characteristics of such a vehicle, designated the 3-Inch Gun Motor Carriage T1, and recommended that a pilot model be manufactured for test.

The basic idea was to produce a light, speedy vehicle for a four man crew. It was to be 14 feet long, 8 feet wide, and 6 feet high, with a maximum weight, fully equipped and with crew, of only 8 to 9 tons. The vehicle was to be powered for maximum speeds of 45 m.p.h. on firm level terrain and of 15 m.p.h. on a 10% grade, and to be able to operate for eight hours at ¾ throttle on one tank of fuel. Mounted at the rear of the vehicle and serviced from the ground, the 3-inch gun was to have a maximum depression of -10° and a maximum elevation of at least 30° and was to be able to go into operation in 39 seconds. Provision was to be made for stowing at least 50 rounds of 3-inch ammunition. Armour was to be limited to a gun shield to protect the crew against frontal small arms fire.

In a memorandum for the Ordnance Technical Committee, the Chief of Infantry objected to the vehicle, stating: "The best antitank weapon is the tank. It alone is capable of employment in counterattack against armored vehicles. Partially armored self-propelled mounts are in the nature of an attempt to appropriate some of the properties of the tank to a defensive vehicle. They have not the offensive powers of the tank. They are not as adaptable to installation in a defensive position as towed cannons. If the vehicle is disabled, the gun is out of action. It is much more expensive than the towed gun and is generally less mobile in cross-country movement."

In spite of these objections, however, formal approval of the project was recorded in January 1941. In March 1941 the Ordnance Committee recorded directives to limit the height to 7 feet, to increase the width to 8 feet, 4 inches, and to fix the time required to go into action at 15 seconds or less.

3" GMC T1 as seen on 21st November 1941. The gunshield has already been added.

Because of various delays, the first pilot 3-Inch Gun Motor Carriage T1 was not delivered at Aberdeen Proving ground until November 1941, by which time the 3-Inch Gun Motor Carriage T24, based on a Medium Tank M3 chassis, had been developed and was available for comparison. The weight of opinion favored the T1 as a strictly antitank weapon because of its low silhouette and its apparent greater maneuverability and speed. The pilot vehicle was built of soft plate and was delivered without a gun shield. It had seats for the driver and assistant driver only and stowage for only 24 rounds of ammunition, carried in open containers in the rear. Because of delays in obtaining the proposed 160 horsepower supercharged Hercules Diesel engine a 90 horsepower Hercules Diesel engine was supplied temporarily.

The vehicle had a travel lock at the front of the gun and a gasoline tank above the sponson on each side. The rear firing platform and the spades for stabilizing the vehicle when firing could be elevated out of the way when travelling. The engine was mounted in the center of the hull, with the final drive and controlled differential in the front. The fans and radiators were in the rear, under the gun mount. Although it was recognized that a number of modifications and improvements were necessary, the vehicle seemed sufficiently promising to warrant standardization to permit immediate planning for manufacture. On the recommendation of the Field Artillery Board and the Chief of Infantry, the Ordnance Committee, on 24 November 1941, gave revised characteristics and recommended that the T1 be standardized as 3-Inch Gun Motor Carriage M5.

Two more views of T1 on the same date

The revised characteristics proposed an increase in the height to 7 feet instead of 6 feet, and of the gross weight to 12 tons instead of 8 to 9 tons. They reduced the maximum elevation to 15° and authorized a change in stowage to 42 rounds. The gun shield was to be of 1-inch armour plate, and a small folding driver's shield was to be provided. The characteristics also directed that sighting equipment for direct laying only be furnished, that pioneer tools be stowed, and if practicable, a radio set SCR-610 be carried.

At this time, discussion was still under way as to the most suitable type of 3-inch gun motor carriage and for a time it appeared that the recommendation would not be approved. However, in January 1942, the Adjutant General's office approved standardization of the M5 and authorized immediate procurement of 1,580 vehicles, subject to approval of final design by the Chief of Field Artillery and the Commanding Officer, Tank Destroyer and Firing Center.

The Brief Automotive Test during November 1941 was discontinued when it became evident that the original 90hp engine was entirely inadequate, but still certain requirements for change were immediately evident.

The gun travel lock constructed at Aberdeen was considered unsatisfactory and replaced by one designed by Cleveland Tractor Company.

The vehicle was originally designed with a seat for the driver and assistant driver in the front portion of the vehicle, and no seats behind the shield for the gunner and assistant gunner/loader. Gunners' seats were then added in a position behind the shield.

The gun as it was originally mounted was provided with a pull-through type of mechanism similar to that employed on the 105mm howitzer. This firing mechanism was very awkward to use as it required approximately 130lbs of pull in order to fire the gun. A modification was thus made by Aberdeen to allow the gun to be fired by a hand lever placed conveniently at the right side of the

Two more views of the original configuration T1, with gunshield.

gunner, this lever had a stroke of approximately 12 and required a force of about 20lbs to operate. It was felt that the mechanism could be improved by the use of a self-cocking mechanism such as on the 3" Gun M7, but it was felt that the necessary changes in the breech block involved too much tooling and at the late date it was felt that the change would delay production too much.

As the spades and firing platform interfered with the vertical obstacle ability of the vehicle and were considered unessential, they were removed. A quick-acting method of blocking out the front bogie and idler suspension springs was employed to stabilize the vehicle while firing.

The vehicle as originally designed provided capacity for 24 rounds of 3" ammunition to be carried in the rear of the vehicle in open containers. It was felt that some closure should be provided to exclude water and dirt, so individual caps were added.

Above: Comparative photos with the T1 in its November 1941 configuration on the left, and on the right in January 1942. The recoil spades are gone, gunners' seats added, and the rear covered, with ammunition caps.

Left: Frontal view of T1 as photographed on 13 January 1942. The fuel tanks are gone, the driver's position slightly modified, headlights added and, most importantly, the engine replaced with a 160hp supercharged diesel. The Cletrac travel lock has also replaced the original effort by Aberdeen's engineers.

Bottom: Left side view of the vehicle also taken in January 1942.

The fuel tanks were relocated in less exposed positions. For positioning the vehicle on targets the gunner communicated with the driver by a speaking tube. A Radio Set, SCR-510, on the right fender was operated by the assistant driver, soon to be re-designated as the gun commander.

The 90 horsepower Hercules diesel engine originally installed as a temporary expedient was unsatisfactory because of lack of power. It was replaced by a 160 horsepower diesel engine on its side, but in

first tests this was unsatisfactory because of overheating of oil. With the addition of an oil cooler and modifications in the cooling outlet to prevent entrained air in the engine block, performance was considered adequate. Some difficulty was experienced with blow-by on the piston rings, but it was felt that this could be overcome by changes in the piston ring design and in the construction of the crankcase heater.

Because of considerable track-throwing and other suspension difficulties, some consideration was given to the advisability of using an entirely different suspension, based on that of the Light Tank M3. However, by the use of straight wing track guides and wide flanges, most of the difficulties were eliminated.

Above: January 1942, the vehicle displaying maximum left traverse and elevation, and also maximum depression. Below left: The new firing handle installed by Aberdeen to the right of the new gunner's seat

The gun used on this vehicle was a modification of the 3-Inch Gun T9, an antiaircraft gun designed to use the same ammunition as the 3-Inch Antiaircraft Gun M3 except for the use of a different fuze with the HE round. For use in the gun motor carriage the guide rails of the gun were lengthened approximately 12 inches and other minor modifications were made to facilitate production in large quantities. Designated the 3-Inch Gun M6, it was used with the Recoil Mechanism M10 and the 3-Inch gun mount M4. Because production of the M5 was limited to 1,500 vehicles, the gun, recoil mechanism and gun mount were designated as substitute standard. Subsequently, Telescope M41 and Telescope Mount M38 were standardized for use with this vehicle, though the carriage started out in life with the unmagnified 37mm sight M6.

In April 1942 it was recommended that the spade be placed back on because there was a question of the accuracy of firing and the rate of fire without it. However, the rear of the vehicle had been modified after the removal of the spade, and it would have been necessary to rebuild it and change the stowage in order to replace the spade, so this was not done.

Left: T1 in January 1942, from a slightly elevated point of view

The original gun mount had the elevating handwheel at the left and the traversing handwheel at the right, requiring operation by two men. The top carriage did not provide sufficiently rigid support for the spring equilibrators. A new mount was designed with both handwheels on the left hand side, permitting one-man operation. This modified mount also provided a more rigid support for the spring equilibrators. These equilibrators were redesigned but were not satisfactory at first.

In order to move the center of gravity forward, an ammunition box for nine additional rounds was mounted on the right fender and armour plate shields were furnished for the driver and assistant driver. At the request of the Field Artillery Board, two more seats were mounted on the front face of the shield, in order that the gunner and loader might ride in a position better protected from dust. Subsequently, the operating seats behind the shield were moved downward and to the rear for a more comfortable operating position and the driver's seat was lowered to a better-protected position.

To replace the individual ammunition caps, doors were placed on the stowage chest at the rear of the vehicle. With the nine additional rounds in the ammunition box on the right fender, the total stowage capacity was 33 rounds, instead of the 42 rounds called for in the military characteristics.

A third travel lock was designed by May, this one proved to be the most successful.

Top: 10th April 1942, the M5 has been fitted with additional frontal armour and ammunition stowage. Expediently mounted in front of the hull are differential oil coolers. Above: By 30th April, the extra seats requested by Ft Bragg were being added.

With the addition of a 1" gun shield, however, the center of gravity was again moved to the rear. This condition was minimized by the reduction in thickness of the front plate of the gun shield from 1 inch to ½ inch. To compensate for its decreased thickness, this front plate was placed at an angle of 35° instead of 18°.

By May 1942, when the modified vehicle was shipped to Fort Bragg for a service test, its gross weight had increased to almost 12 tons, causing a decrease in speed to 36 miles per hour, instead of 38. Addition of the gun shield increased the overall height to 74 inches. The general appearance and characteristics were otherwise unchanged. While at Fort Bragg, the tracks broke twice and the vehicle caught fire. As a result, the proposed tests were not completed when the vehicle was returned to Aberdeen Proving Ground in June.

At a conference in Washington, it was decided to have the manufacturer improve the vehicle to eliminate certain undesirable conditions found at Fort Bragg. Proposed changes included replacement of the engine and increase in the cooling capacity, provision of a revised instrument panel, a change in the third gear ratio, incorporation of a more sturdy track, stiffening of the suspension, addition of improved front armour, and improvement in the methods of firing the gun and signalling the driver. The production type gun mount, with one-man control, was to be installed, a radiator guard added, and a two-power sight incorporated. By 24 July 1942, most of these modifications had been made and six men from the 693rd Tank Destroyer Battalion was brought up from Texas to give the vehicle a service test at Aberdeen. After training the crew, the vehicle was checked for top speed and given a cross-country test, during which it was so damaged that it could not be used for additional testing. The sides were dished in, the gun supports buckled, the suspensions out of line, the traveling lock folded, and the gun mount loosened.

Plans had been made for a comparative demonstration at Aberdeen in early August 1942 with the 3-Inch Gun Motor Carriage T35, based on a Medium Tank M4. The damage to the M5 pilot precluded these cross-country tests, but comparisons were made as to rate and accuracy of fire.

The report dated August 17th made note of the fact that this was still a test vehicle, made of mild steel, not armour, and some relative lack of structural strength was expected, also various strengthening modifications were planned for production vehicles by Cletrac.

Outside of the structural failings, some additional points were noted by the Tank Destroyer personnel. They found the Cletrac to be a generally handier vehicle in avoiding obstacles than the T35 it was tested against. All the cooling problems seemed to have been solved. Fuel consumption was 3.6mpg on roads at 20mph.

These three photographs are dated July 1942. Top left, the M6 gun and mount M4. On the right, the pedestal is added. And above, the mount as it was on the vehicle, visible as the radiator has been removed.

The evaluation was, however, very scathing of the vehicle's vulnerability, especially when compared to the T35, which had more armour, in more places, and for the whole vehicle and crew, and presented a lower silhouette when in defilade.

The M5 on test at Aberdeen, with the Tank Destroyer Board's crew. The driver and radioman do not have their shield's raised, but the gun commander's position actually was to be dismounted to the side of the vehicle. In theory, the gun commander would also have been a radioman and the seat vacated, allowing greater traverse right but they brought a spare crewman along for the tests.

Further condemnation was forthcoming. The firing lever was poorly received as any slight force to the left or right as it was being pulled could throw the gun off target. After every shot, the carriage rocked violently, and combined with the necessity of the blast, both dust and gas, to clear, made re-laying after every shot mandatory; the T35 did not have that problem due to the height of the gun above the ground. "With this gun mount in the Cletrac, fast, accurate shooting under combat conditions is obviously impossible. High standards of direct-laying marksmanship, a basic feature of tank destroyer training and combat, cannot be approached"

The crowded design of the Cletrac also led to difficulty in maintenance functions. Fording ability, at 34" was just over half that of the T35 (60"). Seats were cramped and exposed, and the gun depression inadequate.

The first shot fired broke the lenses on the headlights. The solution to this was a blast deflector which was mounted by August. Colonel Montgomery, TD Board's liaison in Aberdeen, let rip. Excerpts follow:

The Cletrac is pushed to its limit in making 38mph on level smooth roads and is still inadequately protected. Armor has been sacrificed and speed materially greater than good tanks has not been achieved.

An examination of the Cletrac Gun Carriage discloses an extremely compact job of mounting a 3" gun and a diesel motor in a comparatively small tractor. [...] Only 24 rounds have been provided conveniently in the test model. Nine rounds [...] are in an unarmoured box mounted high and exposed on the right track cover. This should be removed. The ammunition provided is too little, whether 24 or 33 rounds. There is no space possible for more on this carriage.

Places for the crew to ride and perform their duties are provided for, apparently according to an after-thought plan. The driver has only a crude and inadequate vision device (hole in shield) much inferior to modern tanks. The same is true of the gun commander. In addition, the commander must dismount, observe, and command from the ground, fully exposed on three sides, his whole situation much inferior to that of the commanders of the enemy's tanks. The places on the gun shield for the gunner and loader must be considered impossible since the men could not occupy thse exposed seats and live in modern combat. They are little better off in the rear seats, even if they could occupy them without serious physical impairment. In these seats, they have practically no view of the enemy to the front.

The maximum of all-round vision for tank destroyer crews is necessary to capitalize on one weakness of tanks - vision, and again, sacrifice of armor is acceptable in exchange for vision. In the Cletrac gun carriage, there is little of either protection or vision.

An instantaneous firing mechanism approaching a 'hair trigger' type is obviously needed in a quick firing gun for direct-laying on tanks which will usually be in motion, seeking concealment, and therefore seen only as fleeting targets. For tank destroyers, every opportunity lost is a minor defeat.

The Cletrac design is fundamentally based on the present band type of track and any necessity of changing to steel track would have unpredictable effects on its speed and durability.

The diesel engine in the tested model is to be installed in only the first 500 Cletrac vehicles manufactured. While it is claimed the change to a gasoline engine will involve only 10% of parts, a serious design problem will enter into the relocation of the fuel tanks which are now directly alongside the engine. It is doubtful that gasoline tanks can remain there, and there is no place to which they can be moved without disturbing the design as a whole.

It is recommended that the Gun Motor Carriage under test not be approved for production because of
a) Failure in basic design to provide space and protection for the necessary crew, ammunition, and dependable radio communication.
b) Failure to operate at speeds required by Tank Destroyer tactical doctrine.
c) Numerous structural weaknesses.

If the Gun Motor Carriage under test, for reasons of expediency, is approved for production and in spite of shortcomings, it is recommended that actual production be delayed until the deficiencies and weaknesses set forth in this report are fully corrected in a production model and opportunity provided the Tank Destroyer Board to give the completed model a full service test. In this connection, it is further recommended that prior to any approval for production with modifications to meet deficiencies, an estimate of the total delay incident to correction of deficiencies be made and this estimate considered in comparison with the production schedule for the 3" Gun Motor Carriage M10 (T35E1).

Below: Fully stowed M5 as on 04 Aug 1942

This page: M5 as photographed 4 AUG 42, with blast deflector added and fully stowed

Because of the rapid progress made in the 3-Inch Gun Motor Carriage T35, using components already tested in medium tanks, it was considered advisable to discontinue development of the M5.

In August 1942, Headquarters, Services of Supply, directed that procurement of the 3-Inch Gun Motor Carriage M5 be cancelled. Authority was given to build two pilot tractors, with the hope that the facilities set up for the 3-Inch Gun Motor Carriage M5, and its proposed companion vehicle, Cargo Carrier T9, might be devoted to manufacture of tractors.

3-Inch Gun Motor Carriage T20

This project was initiated in September 1941. Its purpose was the development of a low silhouette gun motor carriage mounting the 3-inch antiaircraft gun with a muzzle velocity of 2,600 feet per second, and utilizing the chassis of the Light Tank M3 but with a different power plant. To obtain the low silhouette that was especially desirable in a gun motor carriage, a dual installation of such automobile engines as the Cadillac or Packard instead of the original radial engine was contemplated. The design was considered advantageous with respect to both production and maintenance in that virtually all of the components would be production units.

In October 1941, however, at a conference on self-propelled mounts, it was recommended that the project be dropped because the light tank had been judged unsatisfactory for any weapon more powerful than the 75mm gun with a muzzle velocity of 1,950 feet per second. The recommendation was accepted and the project was cancelled that month.

3-Inch Gun Motor Carriage T24

3" Gun Motor Carriage T24 demonstrating maximum official (15 degrees) elevation

Because of anticipated delays in development and production of the 3-Inch Gun Motor Carriages T1 and T20, the Ordnance Committee, in September 1941, recommended mounting a 3-inch gun on Medium Tank M3 chassis. This procedure was adopted in order to save time by utilizing components already in production. The proposed vehicle was designated 3-Inch Gun Motor Carriage T24. The resulting vehicle, it was believed, would be adequately powered, would provide ample room for ammunition and gun crew, and would have excellent armour protection.

In October 1941, the diversion of one 3-Inch antiaircraft Gun M3 and one 3-Inch Gun Carriage M2A2, for use in connection with the development of the 3-Inch Gun Motor Carriage T24 was approved.

The pilot model, built by the Baldwin Locomotive Works, was delivered at Aberdeen Proving Ground early in November 1941.

As delivered, there were a number of interferences noted by Aberdeen, so before subjecting it to any tests, they added a number of stops to limit traverse and elevation. Of note, although the official panel on the pedestal indicated a maximum elevation of 15°, it turned out that there was no stop at all, so a tooth was added to limit the actual elevation to 16½°, but 15° remained the official limit. To protect the recuperator, a stop was added to limit depression to 1° 56'.

Twenty-three rounds were fired in testing, including four as a demonstration for General Barnes. The conclusion was that the mount was stable enough and strong enough to do the job required. However, although the specifications called for a maximum depression of -5°, the gun could be depressed to only some -2° without the aid of jacks, and to -3° with their aid. In addition the silhouette of the carriage and mount was too high and too conspicuous for an anti-tank weapon. The construction of the mount was also considered too elaborate and costly for its purpose as an antitank unit, having a design that did not lend itself readily to mass production. The tests indicated the desirability of a fairly complete redesign, including heavier armour, provision for greater traverse, full protection of the crew against front and flanking fire from small arms, lower silhouette, higher speed, and generally improved performance characteristics.

In January 1942 the vehicle, less gun, was returned to the manufacturer for conversion into the 3-Inch Gun Motor Carriage T40. In March 1942, upon initiation of development of the 3-Inch Gun Motor Carriages T35 and T35E1, it was recommended that the project for 3-Inch Gun Motor Carriage T24 be cancelled. This was approved in April 1942.

Three more views of T24. In the centre (Frontal) photograph, the gun is at maximum depression

3-Inch Gun Motor Carriage T40 (M9)

On 31 December 1941, the Assistant Chief of Staff J-4 approved the manufacture of 50 gun motor carriages based on the Medium Tank M3 chassis mounting the 3-Inch Antiaircraft Gun M1918. Fifty of these guns, the first mobile anti-aircraft gun of its caliber designed and manufactured in the United States, were reported available. The proposed vehicle was designated 3-inch Gun Motor Carriage T40.

As the project for the 3-Inch Gun Motor Carriage T24 had been canceled, it was possible to remove the 3-inch gun M3 from this vehicle and use its Medium Tank M3 chassis for the T40. The tank turret and hull top remained omitted. The same driving controls as on the Medium Tank M3 were utilized but were shifted to the left of the transmission housing.

The gun was mounted on a platform erected directly above the transmission and drive shaft, with the traversing base ring on that platform. In order to cut down the trunnion height, the trunnion stands were cut in half laterally and a one-foot section removed. The two parts were then welded back together again. In doing this, the elevating mechanism had to be redesigned

Top: T40, right side view, 0 elevation. Above: The M1918 at both maximum left and right traverse. Note the elevation handwheel on the right side uncomfortably close to the ammunition box, and the lack of either a sight, sight mount, or port for a sight in the shield. The propeller shaft is visible above the floor, going under the gun platform.

This mounting provided a traverse substantially more right than left of centre. The elevation was limited by a stop on the elevating rack, and depression by the recoil cylinder hitting the front superstructure plate.

The upper two photographs show the 3-inch GMC T40 at 0/0 elevation and traverse, the left profile shows the gun at maximum elevation.

The pilot vehicle was received at Aberdeen on 11 March 1942, and tests begun. As the vehicle was mechanically identical to the well-known M3 medium tank, testing bypassed the automotive components and so the first order of business after measuring was proof firing.

The clearance between the gun breech and the rear of the gun compartment for recoil was 17" minimum. This was at 0 elevation and centre of traverse. The maximum clearance was approximately 25" at maximum left traverse and maximum elevation. The M1918 normally recoilled 40" at 0 elevation. With this in mind, the first firing done was the proof firing of the mount and ajustment of the recoil mechanism. Adjustment of the valve turning arm on the recoil mechanism resulted in 14.5" of recoil at 0 elevation, it was then tapped into place. Nine more rounds were fired at various positions to further proof the mount before testing was stopped: It was obvious that the mount had enough undesirable features to warrant recommendation of redesign of the method of mounting the gun in the vehicle.

The short testing did reveal an number of observations on the carriage as below from the report dated 21MAY42.

The vehicle offers little or no protection for the crew members against weather and also offers no protection from overhead enemy fire.

There is insufficient space between the breech block of the gun and the rear plate to allow full normal recoil of the gun as originally designed. [This lack of space also inhibits] adequate servicing of the gun above 0 degrees and it is impossible to load the gun at elevations above 10 at center traverse.

The elevating mechanism of the gun does not operate freely and is located inconveniently for proper laying of the gun.

Top and centre: Two three-quarter views of T40
Above: The problem of servicing the piece at elevation is clearly shown, as is the lack of recoil space.

There is no sighting device furnished for aiming the gun.

One ammunition box is covered but the other two are not. The box located to the left of the piece and the one located under the firing platform floor are in awkward positions for servicing the gun.

Aberdeen was authorized to modify the pilot model to permit loading the piece at all elevations up to 20° and to relocate the elevating handwheel for one man control. Aberdeen was also authorized to redesign and relocate the ammunition racks.

The matter of the lack of a gunsight was identified early on, with the recommendation on 13th March that Telescope M6 could be used on Telescope Mount T42.

On 30 April 1942, adoption of the T40 as the 3-Inch Gun Motor Carriage M9, Substitute Standard, was recommended in OCM 18143, the original order of 50 vehicles being reduced to 28 when it was learned that only 28 M1918 guns were available.

As recommended for standardization, the M9 was to have a crew of five and a performance comparable to that of the Medium Tank M3 but improved by a decrease in weight. In addition to the 3" gun, it was to have provision for carrying two cal. .45 submachine guns and three cal. .30 carbines.

Top: Rear and frontal views of T40. In the frontal view, the cannon is at maximum depression and right traverse. Above: The rear interior. The cannon is at maximum elevation

The representative of the Tank Destroyer Command filed a nonconcurrence, stating that the vehicle did not meet the tactical requirements of speed, mobility, and light weight which were essential for a gun motor carriage in tank destroyer battalions. He stated, however, that it could be used for limited purposes at the Tank Destroyer Center, pending development of a suitable 3-inch gun motor carriage.

Standardization of the vehicle as 3-Inch Gun Motor Carriage M9 was formally approved in May 1942, OCM 18225, and efforts were made to find a facility to manufacture the additional 27 units. Because so few were to be produced, it was difficult to place the contract, and, in any case, it was believed the cost would be excessive. There was no assurance that all 27 of the remaining guns were serviceable. Accessories, spare liners, and other spare parts for the guns were not on hand and there were no facilities immediately available for manufacture. Furthermore, the vehicle could not be produced prior to the time the 3-Inch Gun Motor Carriage T35E1 would be available in quantity.

Therefore, the Ordnance Department, in July 1942, recommended that the project be cancelled. Formal approval was recorded by the Ordnance Committee on 20th August 1942 in OCM 18689 and the item was removed from the Book of Standards.

3-Inch Gun Motor Carriages T56 and T57

3" Gun Motor Carriage T56

In September 1942, shortly after the standardisation of the Light Tank M3A3 with its improved front hull structure, the Ordnance Committee recommended that this vehicle be used as the basis for two proposed 3-inch gun motor carriages.

The 3-Inch Gun Motor Carriage T56 was to be powered by a Continental W670 engine, similar to that on the light tank, while the 3-Inch Gun Motor Carriage T57 was to be a similar design, but to have the more powerful Wright R975-C1 engine, used on medium tanks.

On both vehicles the turret was to be omitted and the engine compartment moved to the centre, directly behind the driving compartment. The 3-Inch Gun M7 was to be mounted on a pedestal mount at the rear of the vehicle. A cal. .30 machine gun was to be provided for anti-aircraft and ground fire, and provision was to be made for carrying two cal. .30 carbines. A minimum of 40 rounds of 3-inch ammunition was to be carried, as well as small arms ammunition.

Front, side, top, and bottom armour was to be the same as on the Light Tank M3. The rear plate, of ½-inch armour, was to be hinged for use as a loading platform . A gun shield, approximately 1½-inch thick at the front and sides and ½ -inch thick at the top, was to be furnished.

It was believed that the T57, with the Wright R975-C1 engine, would more closely approach the requirements for a 3-inch gun motor carriage as set forth by the Tank Destroyer Command. However, since the T56, with the Continental R670 engine, could be made available much earlier, it was considered desirable to rush this to completion to permit early tests of the mount application.

Formal approval of these recommendations was recorded by the Ordnance Committee in September 1942 and the pilot 3-Inch Gun Motor Carriage T56, manufactured by American Car & Foundry, Berwick Pa, was delivered at Aberdeen Proving Ground in November 1942.

This page: 3" GMC T56 as delivered to Aberdeen, and before the installation of the sight and shield

-*The engine compartment is inaccessible because the ammunition box mounts in front of the engine compartment doors.*
-*General tuning up of the engine is practically impossible without removing the engine from the vehicle. The gun and armor plate shield must first be removed.*
-*The driver's compartment, in front of the vehicle, is too small to afford comfort to the driver and assistant driver, sufficient room for maintenance of transmission oil lines, electrical lines, etc, and room for proper installation of the steering and shifting levers.*

The Continental W670, Series 12 engine, installed in the T56, was a modification of the Continental W670-9A engine, with slightly more power. The maximum brake horsepower was 288 at 2,600 rpm as compared to 250 at 2,400 rpm on the W670-9A. Its torque was 605 lb-ft at 2,200 rpm. as compared to 600 lb-ft at 2,000 r.p.m.

The 3-Inch gun M7 could be elevated from -5° to +25° and could be traversed 15° right and 15° left. The traversing wheel was mounted directly in the rear of the engine compartment and was centered with regard to the sides of the vehicle. The elevating wheel was mounted on the right side of the gun yoke.

Sights were not incorporated in the pilot vehicle, but to facilitate firing tests, a telescopic sight was mounted temporarily on the left side of the gun yoke. A superstructure shield for the crew was locally added as well.

Tests were shortly afterwards conducted, but so many flaws were discovered that the writers of the report ran out of words to enumerate them. Some of the highlights were as follows:

-The maneuverability of the vehicle is poor due to improper weight distribution.

-The rear idler is overloaded tending to over stress the track.

-The effort required for steering is too great.

-The center of gravity of the gun installation is located in the rear of the center of gravity of the vehicle, thereby reducing the vehicle's stability during firing and any vehicle operation.

-One person cannot traverse and elevate the gun due to the arrangement of the elevating and traversing mechanism.

-No provisions for stowage exist in the present design of the vehicle, not considering a limited space in the sponsons and provisions for 40 rounds of 3" ammunition located in the rear of the vehicle.

-The frontal gun shield offers no protection to the crew from the flanks and from the rear.

-No suitable provisions are incorporated in the vehicle for carrying the crew.

-The driver and assistant driver are subjected to unnecessary danger from fire because the gas tanks are located directly behind them.

This page: T56 as photographed during testing, after installation of the sight and locally produced shield.

The traversing mechanism came in for a full page of criticism on its own.

During firing tests, the suspension movement averaged about 14 inches, approximately twice as much as that of other 3-inch gun motor carriages, but the vehicle did come to rest almost exactly where it started. The movement, however, was such that it made rapid, aimed fire difficult.

Automotive tests showed drag-strip performance to the top speed of 39mph being reached in 54.1 seconds, but also demonstrated unacceptable cooling problems, with some parts of the engine hitting 580°F.

The Special Armored Vehicle Board (Palmer Board) recommended that the project be cancelled. It preferred the 75mm Gun Motor Carriage T67, which it also tested at this time, and wanted all the other experimental projects dropped in order to concentrate on it.

Aberdeen Proving ground, in its formal report, concluded that the vehicle, as tested, was unsatisfactory as a tank destroyer. To make it suitable it would be necessary to lengthen the suspension, widen the tracks, enlarge the crew compartment, redesign the engine compartment, improve the gun installation, and make several other changes. Such a total redesign, however, would introduce the possibility of overloading the suspension and of making the vehicle too heavy for its engine, the report stated.

The 3-Inch Gun Motor Carriage T57 was delivered at Aberdeen proving ground in December 1942, shortly after the Special Armored Vehicle Board had submitted its unfavorable report on the T56 (See below), with 119 miles on the clock. Steering was easier than on the T56 because the control levers were 3 inches longer and were placed two inches farther forward.

3" Gun Motor Carriage T57. Note the external air filter on the left rear as an obvious identification feature

The Wright R975-C1 engine provided greater power. There was a difference in the transmission adapters, as well. The T56 had a 1:1 ratio, while T57 had a 1:0.743 ratio, reducing the torque input into the final drive 26% and increasing the speed also by 26%. No gun shield was furnished, and this time Aberdeen did not mount one. In other respects, the vehicle was similar to the T56, and was subject to the same criticisms.

Overall, T57 was tested from 17 December 1942 to 13 April 1943, during which the vehicle was operated a total of 887 miles; 107 miles of which were cross country, 150 miles on secondary roads, and 510 miles paved. The full series of tests was not completed, however.

There was concern that with the increased power of the engine, the greater leverage from the new steering levers, and higher speeds, the standard light tank brake bands would not prove strong enough. Sure enough, one snapped and the other wore heavily, but the evidence showed that the cause was a faulty transmission and clogged oil pump which exposed them to excessively high temperature and insufficient lubriction. A replacement of these components and the brake bands seemed to fix the problem.

The problems with engine accessibility were no better with the new engine. The oil filling port was in the far corner of the compartment, and could only be filled by use of a pipe and quart container. It took several hours to fill the oil resevoir. Taking the engine out for almost any routine maintenance operation also remained a problem.

Track tensioning required lifting the rear of the vehicle with jacks to reduce the weight. However, doing this also made it impossible to determine if the adjustments were correct. Tensioning track thus became a very time-consuming event.

A more minor issue was that the driver kept hitting his elbow against the rear wall of his compartment when shifting into low, third and fifth gears.

The new engine did help. The vehicle clocked nearly 50mph at 2,200rpm on hard surface, and 30mph on dirt.

3" GMC T57 at max elevation

3" GMC T57 with gun in travel lock

The Special Armored Vehicle Board in the meantime, on December 3rd, issued its own opinion of the programs of the Rough Pilot Model Tank Destroyer T56 (and by association, T57), and its general obsevations mirrored those of the Aberdeen Proving Ground test report.

In the opinion of this Board a gun motor carriage for tank destroyer use must be a self-contained unit capable of carrying the gun crew and all necessary ammunition and equipment without recourse to an accompanying vehicle. It is not believed that so large a gun as a 3" or 76mm can be so mounted on a light tank chassis as to provide either a stable firing platform or suitable characteristics just referred to.

None of the using services represented on this Board (The Armored Force, Tank Destroyer Command, and The Cavalry) desires Gun Motor Carriage T-56.

This Board finds no reason for further consideration of Gun Motor Carriage T-56 "for service use or for further use or for further service test or to outline further developments to be pursued" by the United States Army.

The Board unanimously recommends termination of further development of the Gun Motor Carriage T-56 for the US Army.

Because of the adverse report of the Special Armored Vehicle Board on the T56, the Ordnance Committee in February 1943, recommended that the projects for the 3-Inch gun motor carriages T56 and T57 be cancelled and that the vehicles be retained at Aberdeen Proving Ground for use as proof facilities and for reference purposes. OCM 20069 of 24th February 1943 cancelled the project, and the proposed tests of the T57 were not completed.

3-Inch Gun Motor Carriages T35, T35E1 (M10)

3" Gun Motor Carriage M10

As a common, in-service vehicle, the M10 has been covered in great detail by other authors in the past. The development history is covered below for the purposes of completeness of this volume, but little attempt will be made to cover the vehicle's in-service history. M36 and M18 shall be treated similarly.

By November 1941, as the pilot 3-Inch Gun Motor Carriage T1 approached completion, considerable opinion developed in favor of a heavier vehicle, with better armour protection and more adequate stowage. Tests of the 3-Inch Gun Motor Carriage T24 indicated the possibilities of a vehicle based on a medium tank chassis and pointed the way toward an improved gun motor carriage. The T1, because of its lower silhouette and apparent greater maneuverability and speed, was favored over the T24 and was recommended for standardization as the M5, but almost simultaneously the Ordnance Committee recommended development of a new vehicle, designated 3-Inch Gun Motor Carriage T35.

Like the T24, the T35 was to be based on a medium tank chassis. Instead of the Medium Tank M3, however, its basis was the Medium Tank M4A2, powered by dual General Motors diesel engines, which was just starting to come off production lines. It was to mount the 3-inch gun as modified for Heavy Tank T1, in a 360° turret similar to that of the heavy tank except that it was to be thinned down to reduce its weight, was to be open at the top, and was to omit the coaxial cal. .30 machine gun.

While this project was still in the design stage, reports from the early fighting in the Philippines highlighted the disadvantages of vertical armour, as used on Light Tank M3. These reports pointed out that sloping armour was better able to deflect enemy projectiles. In accordance with "anxious" requests from Colonels Sawbridge and Negrotto of the Tank Destroyer Board for a vehicle with sloping armour and lower silhouette, the Ordnance Department hastily worked out three designs, which differed mainly in their silhouettes, with overall heights of 89", 94" and 96".

One of these, with a 94-inch overall height, was approved as an alternate design to the T35 as the 89" was deemed to provide insufficient visibility, and was designated 3-Inch Gun Motor Carriage T35E1. In January 1942 the Fisher Tank Division started work on final design drawings of both the T35 and T35E1 and by April 1942 had completed a pilot model of each.

Left: The rather basic mockup of T35, photographed 20 January 1942 with the originally proposed turret

T35

The project was initiated with Ordnance Office letter 451.25/13699 dated 25th November 1941 and drawings of a proposed gun motor carriage utilizing the Medium Tank M4 hull with 1" armour basis instead of 1½" and the Medium Tank M4 turret, without cage, having the rear portion cut out in what the official record describes as "the manner of a Roman chariot." Quite how this was to be balanced is not found in the record. The 3" gun T12 was to be mounted in the turret similar to the Heavy Tank M6 mount. The motive power was specified to be "General Motors Dual 6-71 engine modified so that a top speed of 35mph in high gear may be obtained providing weight permits."

With the exception of the turret, very little design work was done on the T35 as it was substantially a Medium Tank M4 with reduced armour thickness. The cal. .30 bow gun was removed as the Tank Destroyer Board felt it was unnecessary. Auxiliary armour plate, set at a 45° angle, was bolted over the three-piece lower front plate. As on the medium tank, entrance hatches and direct vision slots for the driver and assistant driver were set in the sloping upper front plate.

3" GMC T35, showing well the angled front plate in front of the transmission housing, and the original design of the turret for the T35 series. Photograph taken 23APR42

In order to get better turret ballistics and yet retain the Medium Tank M4 ring the sides were bulged to form a 33 and 31 degree angle with the vertical. The Roman chariot concept was soon dropped, partially for more protection, and partially also because it seemed that nobody else understood how the thing was supposed to be balanced either. Even with this added weight in the rear, the turret center of gravity fell over 12 inches ahead of the ring center line and caused traversing difficulty on sloping ground. A ¾" spacer was required for T35 to clear the M4's ring; this was not to prove necessary on T35E1.

The gunshield as seen on T35

The protection for the gun was improved by adding a 45 degree V shield over the slot in the turret front plate. The addition of this shield eliminated the costly machining of the rotor and turret front plate. The turret traverse was based on the Medium Tank M4 traverse and had a ratio of 58 to 1. The elevating mechanism consisted of a pivoted screw hanging from the turret roof with a rotating nut fastened on the gun cradle. This nut and screw mechanism required no expensive parts and could be made on machinery readily available. The heavy tank sighting system was adopted, but more room was allowed between the telescope and the right recoil cylinder.

In the meantime, testing had been conduced with a modified Medium Tank M3A3 in late Decmber. The tank had the twin diesel engines combined with a new set of 1.6:1 reduction gears to see whether or not they actually could reach the requested top speed of 35mph, and to see if the track and suspension could actually withstand it. It turned out that the testing was a success, the tank topped out at 37mph, and it was felt that a slight reduction in the gear ratio could be permitted to have the engine top out at higher revs while dropping the speed 2mph. This would allow better acceleration in reduced traction conditions such as mud.

Left side view of 3" GMC T35, also 23APR42

The T35 was finally given a quick test on the pavement outside the shop, then loaded onto a flatcar, without gun, to be sent to Aberdeen, where it arrived on 16 April.

When Aberdeen received it, the first thing they did was to take a 3" AA Gun T12 from a Heavy Tank T1, and install it into T35. The log indicated it was a simple process, and that after the counterweights were installed, the balance was not as far forward as it was in the heavy tank, even without the 37mm. After a quick familiarisation, some changes were immediately deemed necessary. The elevating mechanism was deemed to have been constructed without close enough tolerances, and the handle needed to be relocated due to interferences with the traverse gearbox at extreme elevations. The traverse ratio of 58:1 was very quickly deemed unsatisfactory as being too coarse for fine aiming, and not providing enough torque to turn the unbalanced turret on a slope. A standard M4 traverse mechanism, 400:1, was installed.

Three more views of 3" GMC T35

Initial observation was that the absence of a turret basket was not a particular disadvantage. The position of the loader was very close to the center of the turret, and he seemed to be able to perform at all points of traverse. The recoil guard needed a little tuning for shape, but the mechanical linkage for firing was deemed to require too much leverage; some modifications would have to be made. "It is definitely contemplated that the firing will be by mechanical linkage in order to keep the turret free from slip rings and wiring as the overall tenor of the design is to keep the whole unit as simple as possible". The elevating mechanism as well was concluded to be too coarse, and changed from 570:1 to 1,700:1

Above, below: The turret on T35 was obviously fitted out only to the miniums required for concept testing.

Finally, the vehicle was ready to be put through its paces, with a first run on 24 April 1942. "The performance was gratifying with evidence of plenty of reserve power." The vehicle had been governed to 30mph, and averaged 27.8 over the two mile course, noted favourably against the 3" GMC T1 which averaged 32mph despite being governed to 42mph. Ease of driving was rated as "even easier than an M4 tank." However, it was also noted that a very sturdy travel lock would have to be provided, as the turret's unbalance would set up considerable turning moments. Imbalance on the gun was partially addressed by the implementation of new bushings, and sleeve bearings really only changed the periods of oscillation.

When T35E1 arrived a few days later, the vehicles were demonstrated on 2 May 1942. At a meeting afterwards, the decision was reached to standardise the T35E1 with certain weight modifications. The project for the T35 was cancelled, but the pilot vehicle retained at Aberdeen for tests of design features identical with those of T35E1, and for possible use as a training vehicle.

Below left: T35's turret after ballistic testing with 37mm ammunition.
Below right: T35 GMC, front view.

T35E1 (M10)

The 3-Inch Gun Motor Carriage T35E1 had the same lower hull, but the side and upper rear plates, instead of being vertical, were sloped in sharply toward the top. Furthermore, the sides and rear had an armour plate skirt, ½-inch thick, which sloped in toward the bottom, giving additional protection to the top of the tracks and to the lower hull.

The front upper plate was made in one continuous sloping piece, the direct vision slots being omitted and the entrance hatches for the driver and assistant driver being set flush in the hull deck. The turret sides and rear were bulged to form a 35° angle with the vertical.

When the first pilot vehicles were received at Aberdeen Proving Ground it was specified that the gun to be used on both vehicles should be the 3-Inch Gun T12, later standardized as the M7. This gun, also used on the Heavy Tank M6, had interior dimensions and ballistics practically identical to those of the 3-Inch Gun M6, used on the 3-Inch Gun Motor Carriage M5. It utilized the same trunnion assembly and bearings as the heavy tank.

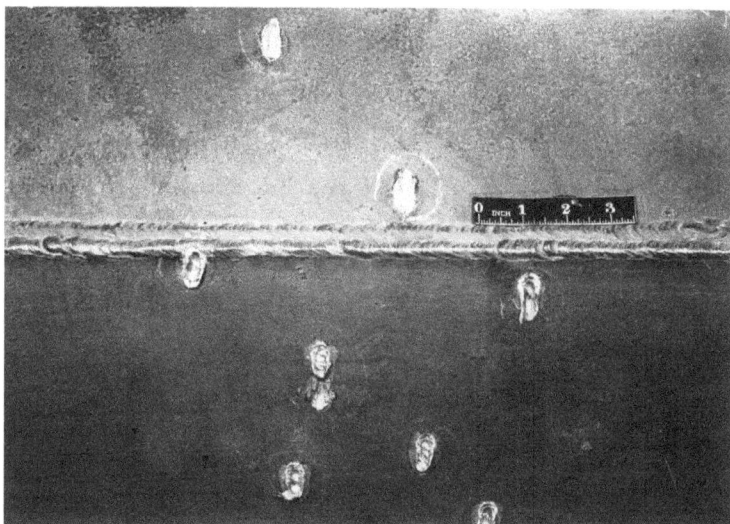

Testing of the angled surfaces with cal. .30 ammunition

Firing tests against both the T35 and T35E1 with cal. .30 and cal. .50 AP and ball ammunition gave a striking demonstration of the greater protection afforded by the sloping armour of the latter vehicle. It appeared from the firing that an angle of obliquity of 45° or better was necessary for the particular rolled plate used in order to ricochet the bullets. At 38° the bullets would gouge in deeply before being deflected, whereas at 45° and 54° practically no penetration occurred. The armour plate skirts of the T35E1 did an efficient job of turning bullets up to cal. .50, preventing them from injuring the tracks. The firing tests also showed that the cast turret was much inferior ballistically to the rolled plate.

As approved for standardization the weight of the 3-Inch Gun Motor Carriage M10 was to be reduced and a welded turret was to be used instead of the cast turret. In order to reduce the weight to 58,000 pounds instead of 61,000 pounds, the hull armour was reduced in thickness, that on the sides to ¾" instead of 1"; the skirts to ¼" instead of ½"; the front plate to 1½" instead of 2" and the engine deck to ⅜" instead of ½". The lower hull was left at 1". As the Proving Ground felt that even this reduction in armour protection was a drastic sacrifice for the weight saved, bosses were welded to the front and sides of the hull and to the sides of the turret, to which extra armour plate could be bolted if required. British documents indicate the thickness of the plates to have been 17mm, but officially they could be in varying thicknesses. Testing in Aberdeen between September 1943 and March 1944 involved plates from ¼" through 1" at varying angles and spacings up to 8": The conclusion was that no spacing at all was best.

In designing the welded turret, an attempt was made to arrive at the most desirable turret shape to house the 3-inch gun from the standpoint of operating convenience and maximum protection while facing enemy fire. A radical deviation was made in the gun mounting and the gun rotor and turret front plate were eliminated entirely, being replaced by the gun shield. In order to get an installation that could be readily removed in the field, the customary gun trunnion construction was inverted and the trunnion pins were placed in the turret with the trunnion bearings a part of the gun yoke. The trunnion pins were made removable in order that the 105-mm howitzer or the British 17-pounder gun could be interchanged with the 3-inch gun, if desirable. However, such installations were never made by the US.

This page and next: Various views of T35E1, taken 04 May 1942

The trunnion bearings incorporated bronze sleeves instead of roller bearings to introduce a friction factor to damp out the bobbing of the gun with an out-of-balance condition. The elevating screw was moved from the right side of the gun to a position between the left recoil cylinder and the gun tube. This permitted stronger construction and allowed for dual elevating handwheels and opened the way for complete dual controls of the turret.

Aberdeen Proving Ground had originally designed and mocked-up a turret of a hexagonal shape, consisting of five sides and the gun shield. A number of problems arose and a pentagonal-shaped turret, consisting of four welded sides and the gun shield, was used as a temporary expedient. In July 1942 it was decided to revert to the original shape, but by this time tools and jigs had been ordered. As a change would have delayed production, the pentagonal-shaped turret continued to be used.

Upon completion of the first production model of the M10, Aberdeen Proving Ground reported that it embodied excellent mobility, provision for auxiliary armour, good ballistic qualities and that it presented a minimum of mechanical complications. The Proving Ground considered sloping armour more desirable than vertical armour because it gave greater protection for a given weight. As the design embodied straight line welding it would lend itself to automatic welding and efficient production. Aberdeen felt that the vehicle was well-balanced for fighting at long ranges and that it would be best able to receive enemy fire from the front or front quarter. The open top turret allowed good vision, good ventilation, and good lighting.

The Proving Ground recommended that work be initiated on a power traversing device for the turret to increase its ease of handling and that dual controls be incorporated as soon as possible. It suggested that the assistant driver's extra periscope be eliminated.

The production vehicles employed a newly-designed one-piece lower front plate. This eliminated the need for the 45° V auxiliary armour which had been used to protect the three-piece lower front plate on the pilot vehicle. Production plans were under way a few weeks after standardization of the vehicle.

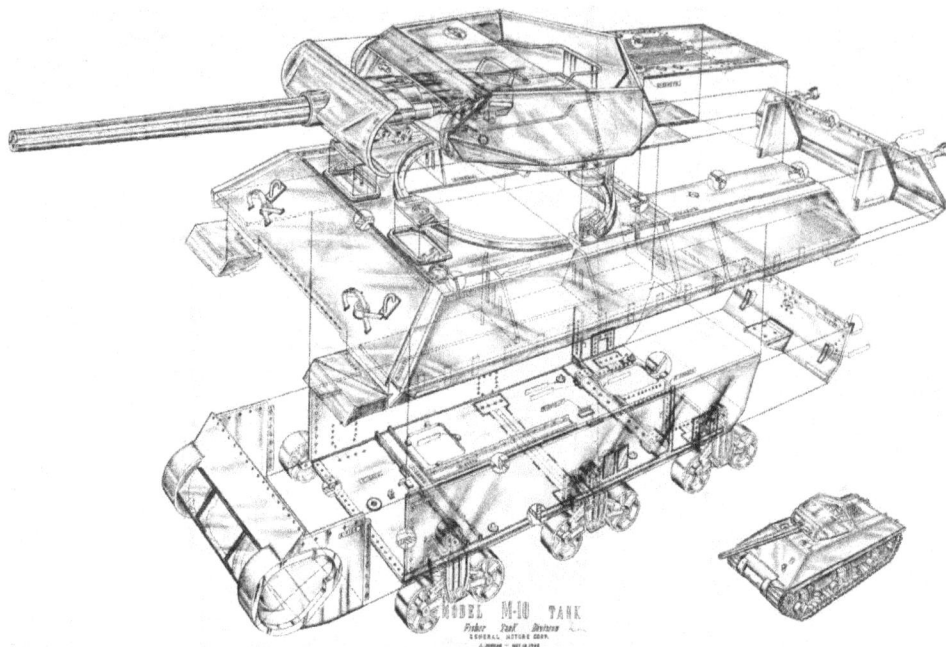

It received an AA1 priority rating, which was even higher than that of the medium tanks then in production. To speed manufacture, the design was frozen except for minor details. In order to bring production to a point which was considered ample, arrangements were made to have the Ford Motor Co. manufacture a similar vehicle, based on the Medium Tank M4A3 which used the Ford GAA gasoline engine with a different transfer case gear ratio (1.37:1) to give the same propeller shaft speeds. This vehicle was designated 3-Inch Gun Motor Carriage M10A1.

By 30 September 1942, 105 vehicles had been completed and by the end of the year they were being produced in quantity and rushed overseas. Because of their similarity to the medium tanks they could be assembled in parallel production lines, presenting the unusual situation of tanks and tank destroyers issuing from the same factories.

Top: Exploded diagram of M10 by GM. Above: M10 at Aberdeen, 9 Sep 1942

In a report dated 19 December 1942, the Tank Destroyer Board, Camp Hood, Texas, reported that the M10 was superior to any of the other expedient tank destroyers then available (37mm GMC M6 and 75mm GMC M3). The Board stated, however, that the vehicle did not comply with the military characteristics of a "highly mobile 3-inch gun" as originally established, since it had only a moderate increase in speed over existing medium tanks. Because of this, the vehicle had no tatical advantage over medium tanks except in firepower and this advantage could be fully utilized only with a higher mobility than that of a tank, the Board declared. For the curious, they timed how long it took to fill the fuel tanks using 5-gallon cans: One hour.

Internal view of an extremely clean M10

The TDB's wish list primarily consisted of:

     a. Counter-balancing the turret (Tracking at cant angles over 4.5° was difficult, and impossible at 6).

     b. An improved sight mount and sight for direct laying.

     c. A simplified sighting system for indirect laying.

     d. Power traverse, by installing an additional generator if required.

     e. A rugged mounting for the traversing handwheel, or, if power traverse could not be supplied, a second handwheel. The latter especially true if a counter-balance is not fitted.

"Deadeye" was the M10 which found its way to Fort Knox for testing by Armored Board

Armored Force did its own testing of the 3" Gun Motor Carriage M10 at the end of 1942, with its short report being published 29 January. It conducted a comparison against the 75mm M4 tank, and some of the criticisms are interesting considering Armored Force were at the same time happy to see a 76mm gun on M4.

(a) The 3" gun, M7, has little advantage over the 75mm tank gun, M3.

     1. It is bulkier and requires more room to service.

     2. The increased size of the ammunition limits the number of rounds carried.

     3. Because of the muzzle blast it is impossible for the gunner to sense his strike or use tracer control at ranges under 1500 yards.

     4. It is no more accurate for fire at stationary targets at practical ranges. See Appendix "E" (Not included in the copy the author found)

     5. It is very difficult to remove and clean the breech block. The firing mechanism cannot be replaced without damage to the trigger plunger unless a special tool is provided.

     6. The 3" gun, M7, will penetrate greater armor thickness

	1,000 yds	*1,500yds*
3" Gun M7	*4.1"*	*3.6"*
75mm tank gun, M3	*3.6"*	*2.8"*

(Shot APC against homogenous plate ; maximum angle of incidence, 20 degrees)

(b) The trajectory of the 3" gun, M7, is too flat to make indirect fire practical. Therefore, the panoramic sight, mount, and linkage are unnecessary.

(d) The positioning of the eyepiece of the telescopic sight is so far forward and the oversized traversing handwheel is so far to the rear that accurate fire control by the gunner is very awkward.

(e) The traverse of the antiaircraft machine gun is 1imited to 120° and is inadequate. The slow speed of turret traverse minimizes the possibility of extending this field by rotating the turret.

(g) There is inadequate protection against enemy infantry. A bow machine gun should be installed in space now available.

(h) The gun carriage is stable in all firing positions.

The comment on indirect fire is notable, as the Tank Destroyer branch made indirect fire standard training for its gun crews, and it was a common role for them in theatre. That said, the final report on M10 by the TD Board made a similar comment: "The flat trajectory characteristic of the gun is unsuited to indirect fire." However, they merely recommended a simpler indirect laying system.

Left rear of the M10 in Aberdeen Proving Grounds, Sept 42

Overall,

a. The Armored Force Board concludes that that The 3" Gun Motor Carriage, M-10, corresponds to the Medium Tank M4 series, in mobility and mechanical efficiency.

b. The 3" Gun Motor Carriage, M-10, has inadequate armor protection for tactical employment in the Armored Force.

c. The 3" Gun, M7, has little advantage over the 75mm Tank Gun, M3.

d. The unbalanced turret and inadequate traversing facilities greatly hinder the effective employment of the 3" gun.

RECOMMENDATIONS: The Armored Force Board Recommends that the 3" Gun Motor Carriage, M-10, be considered unsuitable for use by the Armored Force.

As an aside, the Armored Force Board report was consistent in its use of a dash in the nomenclature "M-10", whilst being equally consistent in the lack of one in "M4", "M7", etc.

In the meantime, a number of minor changes were made and efforts were made to overcome what were considered major problems. Both Army Ground Forces and the Tank Destroyer Board echoed the recommendation of Aberdeen Proving Ground that a power-traversing mechanism be provided for the turret. Development of a suitable dual control for the turret traversing mechanism was also initiated.

From the beginning, the use of the heavy 3-inch gun without a compensating bulge on the rear end of the turret had thrown the turret out of balance, causing difficulty in traversing on slopes and in holding the turret in the locked position while negotiating cross country terrain. On the first vehicles, this was partially overcome by mounting the antiaircraft gun on the rear corner of the turret, hanging grousers on the rear, and stowing ammunition on the rear. Later, counterweights were mounted at the rear of the turret of production vehicles.

When the M10 was standardized, the military characteristics specified the use of fire control for direct fire only. Subsequently, however, the Commanding General of Army Ground Forces informally recommended that provision be made for both direct and indirect fire and that magnification of the telescope for direct laying be increased to three power. Further investigation by the Tank Destroyer Board indicated that it was advisable to eliminate the entire mechanism proposed for indirect laying and to provide in its stead a simple device geared to the turret ring for setting off horizontal angles. This indirect laying mechanism was to be replaced by an azimuth indicator modified from that used on the Medium Tank M4, but until this could be produced, it was agreed that the vehicle would be provided with direct fire sighting equipment only. Accordingly, in March 1943 the Ordnance Committee recommended that fire control equipment consist of a Telescope M51, a Telescope M59, a Telescope M44, three Telescopes M6, an open sight, an Azimuth Indicator M18, and a Gunner's Quadrant M1. This recommendation, which also applied to the 3-Inch Gun Motor Carriage M10A1, was approved in May 1943.

In April 1943, the Ordnance Committee recommended that a spur gear manual traverse mechanism be built to replace the production worm gear manual traverse. The new device was expected to have greater efficiency, permitting operation of the turret on steeper grades without the use of as much balance weight as was formerly required. It was to have a safety clutch between the drive pinion and gear train to prevent the breakage to which the production unit was subject because of the shock imposed by the unbalance of the turret when the vehicle operated over rough ground. The design was to permit individual or dual operation as desired. Actually, two mechanisms were to be developed; one with a single speed range for immediate field replacement of the production mechanism, and the other with two speed ranges for possible future use. The Ordnance Committee recommended that four pilot models of the single speed traverse mechanism be obtained for installation and test. These recommendations were approved in May 1943, and work was begun on the project. A single-speed, single station, spur gear type box was designed to meet immediate field requirements. Four pilot models were built and one was tested for performance and durability, with excellent results. However, this design was abandoned since there was interference between the handle of the traverse mechanism and the M51 telescope mount, when the gun was elevated to 17° or higher. A new design covering a two-station mechanism built around a differential type gear box was begun.

133

During the period when the dual hand traverse was being developed, the Oilgear hydraulic power traverse was being used with great success on medium tanks. A 3-Inch Gun Motor Carriage equipped with one of these units was shipped to the Tank Destroyer Board for testing. In December 1943 the Tank Destroyer Board reported that the test had very favorable results, and that the power traverse was far superior to the hand traversing mechanism. However, this project, as well as that for the development of the dual hand traverse, was later cancelled, because the 3-Inch Gun Motor Carriage M10 was going out of production.

Tank Destroyer Board did attempt a few other modifications. One was a modification to the exhaust system, as it was felt that the M10 raised so much dust that not only was it a hindrance to concealed movement, but that it was also a nuisance to following vehicles. The Training Brigade also noted that if the M10 towed a 1-ton ammunition trailer, the heat and dust tended to have deterious effects on the trailer and its cargo, but the trailer was so hated that the brigade flat refused to tow it any more anyway. Tank Destroyer Board also tried installing a ring mount for the .50 cal. Even the counterweight needed modification, to stop the radio antenna from getting caught in the gap between the two halves: The solution was to weld a small metal bar across the gap.

Testing of M10A1 continued through the end of 1943 by the Tank Destroyer Board. A report dated 12 February 1944 gave a comparison between M10 and M10A1 (and to a limited extent T70) during a 2,000 mile endurance test.

The M10 only had one significant advantage: Fuel and lubricants economy. Over the course of the runs, the M10's diesels would drink a gallon over 0.978 miles, and also burn a quart of oil after 152 miles. The Ford V8, however, would only do 0.66 miles to the gallon, whilst using a

Top: Proposed modification to the exhaust system as implemented in Camp Hood to reduce dust signature.
Above: the ring mount for the cal. .50 HMG as tried by TD Board

quart of oil every 20.3 miles. In addition, the average speed over the 2,000 miles was 1 mph less for the M10A1. However, there was more mud during the M10A1's run. The T70, in comparison, ran 0.74mpg, at 8.1 mpq of oil. However, the higher horsepower of the M10A1 gave "better acceleration and performance in high gear than the M10, repeatedly commented on by drivers familiar with both."

The report concluded:

a. The Ford engine, Model GAA V-8, is superior to the GM Diesel, Series 71, Model 6046, twin-six, in the M10 type Tank Destroyer.

b. No important difference in tactical mobility of M10 and M10A1.

c. Steel tracks are more destructive to bogie wheels and idlers than rubber track.

[...]

f. Bells or buzzers are needed in the circuits for warning lights on the instrument panel. (This last to attract the driver's attention when driving head-out on road marches.)

The report thus recommended that the M10A1 be considered the superior to the M10 for tank destroyer use.

The 3-Inch Gun Motor Carriage M10A1 was not sent overseas with the US Army because by the time it went into production sufficient M10s had been delivered to meet requirements. A number of these were converted into tank recovery vehicles and others into Full-track Prime Movers M35, used for towing 240-mm howitzer and 8-inch gun materiel. And, of course, into M36s, see later entry.

Progress on the 76mm GMC T70 (again, see later entry) indicated the desirability of replacing the 3" gun in tank destroyers with the 76mm.

A letter from General Bruce to General McNair in January of 1943 expressed the desire to create a common turret for T70 and M10, which used the same turret ring, to replace the unsatisfactory M10 turret which required 3,600lbs of dead weight just to balance it, whilst giving good testing to the turret to be placed into T70. In addition, MG Bruce observed, it would be desireable to reduce the requirement to produce and develop new 3" ammunition. (To this end, a towed 76mm was being developed to replace the 3" Gun M5 which saw service with TD units).

In this letter, MG Bruce suggested that, unless a suitable turret was not likely to be coming soon, the M10s be produced as hulls only, to allow the new turret to be dropped in.

For more on the story of the 76mm-armed M10, see the entry for T72.

Two views of the T70 turret on M10

Nothing much seems to have come of this letter. The next entry in the file in the Archives is a letter dated 22 May 1943, from LTC W.E. Sherwood, the Tank Destroyer Center's liaison officer to the Tank-Automotive Center in Michigan, in which he states that nobody he had talked to had seen the letter. This, thus, was part of the reason that no action was apparently taken. However, the Tank-Automotive Center also gave an official reason as not wanting to interfere with Buick's construction of T70 pilots. LTC Sherwood opined in the letter, though, "other reasons may have contributed to the reluctance to try out this idea," which may have been in relation to T-AC's apparent interest in the T72 program.

In the end, the Developments Branch found an old shop model of a T70 turret, absent gun and interior fittings, and on about 10th May 1943 placed it onto an M10 chassis in Grand Blanc, Mi. Quite what this experiment was to show is unclear, as LTC Sherwood indicated that he was to understand that a properly functioning turret from the third T70 pilot was to be placed onto an M10 in Aberdeen Proving Grounds to create a better example, and only so much could be learned by noting that the T70's turret happened to fit into the same sized turret ring on M10. Still, the project did provide interesting photographs to confuse future AFV enthusiasts.

Side view of the above T70-on-M10 experiment

Finally, a note on names. The author has not seen any official documentation assigning a name to the M10 (outside of the British designation of "Achilles" in spring 1945), and in particular, nothing for "Wolverine." David Fletcher, of Bovington, is quite certain that it was not a British name, and it does not fit in with their naming conventions of the time anyway. Similarly nothing has been found in the US archives. Although the Canadians were known to use names of animals for their vehicles, investigations through the Canadian War Museum resulted with a statement that they did not believe the name was of Canadian origin. It is, of course, impossible to prove a negative of this sort, but until evidence is available to the contrary, the author's conclusion is that any name is a post-war retroaction of non-official origin.

76mm

The 76mm was designed to fire the same projectiles as the 3" gun, but at much lighter weight. This is the gun that ended up arming the M4 medium tank, as well as the M18.

76mm Gun Motor Carriage T70 (M18)

76mm Gun Motor Carriage T70 Pilot #3

The development of the 76mm Gun Motor Carriage M18 marked the end of the search by the Tank Destroyer Command for an ideal tank destroyer combining speed, mobility, and striking power greater than that available in tanks. A set of specifications for an ideal vehicle of this type was submitted in February 1942, and from these was developed the 57mm Gun Motor Carriage T49, based on the 37mm Gun Motor Carriage T42. As described earlier, from the T49 evolved 75mm Gun Motor Carriage T67, based on the T49 chassis, but mounting an even more powerful gun. The favorable performance of this vehicle throughout tests resulted in a recommendation by the Board that this vehicle, with a standard engine and other changes, be developed with a view toward early standardization. In accordance with this recommendation and with the desire of the Tank Destroyer Center for the more powerful 76mm gun, a study was begun to incorporate this gun on the T67 chassis. By early December 1942, it had been concluded that the gun on T67 could be changed to a 76mm very simply.

This final vehicle, which was designated 76mm Gun Motor Carriage T70, was to have a 900T torqmatic transmission with multiple disc clutches for first and reverse gear, in place of the 300 torqmatic transmission with band clutches used in the T67. The weight of the heavier 76mm gun, concentrated in the rear of the vehicle, made necessary a revision of turret and power train design. The turret was balanced by addition of a large bulge in the rear end, and the transmission was moved to the front.

The poor performance of the Buick engine in the T67 led to inclusion of a power train similar to that of the Medium Tank M7, using the Continental R975-C1 air-cooled, radial engine. In part of the selection process, Buick engineers had been instructed to consider not only this engine, but also the Ford GAA, the Cadillac light tank engine, and the Continental 6572, with production to have priority over performance. A record by Colonel Haskel of TD Branch of a telephone call with Colonel Colby, at the Tank Automotive Centre on 11 Dec 1942 states "The R-975 was only considered because of its availability and the fact that it may be the only engine suitable as a result of engineering studies." The record indicated a preference for the Cadillac engine if possible, though, with changes to the vehicle it would come down to 13 tons with a hp/weight ratio of 20/ton. This same call also instructed the design of the vehicle with torsion bar suspension, and Colonel Colby expressed "he was glad that we had selected the 76mm as he feels it is possible to encounter tanks with great armor protection which would need the extra punch developed by this gun".

In the end, it came down to the Wright and Ford engines. The Wright required the change to a front sprocket configuration, using the M7's controlled differential, and also a hump in the hull rear which limited depression in that arc and also raised the silhouette of the vehicle a little. On the plus side, the move meant that the vehicle could be 5" shorter. It was anticipated that in the long run the radial would be withdrawn from use, and replaced in production by the Ford V8, but there was some abundance of caution in TD Branch at this point and it was decided to wait and see if the Armored Force decided that they were satisfied with the engine in the tanks before installing it into their premier TD. The actual installation was considered to be a very simple prospect. The inclusion of a bustle on the turret was deemed a necessary evil, as the gun was just too heavy to have the turret balanced without one, and the trunnions could not be placed further back.

In a conference on 12 Dec 1942, Buick was "flatly asked" two questions: Could 75mm T49 go into production as is, and did Buick think they should go with that or the new 76mm T67? (Bear in mind, again, that the designations did not keep pace with the changes in armament and configuration). The answer was that T49 was indeed viable for production, but with the 75mm only; The 76mm was just too heavy to keep the carriage balanced with the turret and gun's center of gravity where it was. Buick recommended going with the new vehicle: As most of the components were already known and in production, unlike for T49, getting T67's bugs worked out and the vehicle into production would take about the same amount of time as getting T49 into production, and T67 seemed to be overall a far more promising vehicle. It was hoped that six pilot models could be produced without delay.

As an aside, a note from Tank Destroyer Command's liaison officer in Detroit on 29DEC indicated "It is likely that the designation T67 will be abandoned when the OCM passes on this new vehicle and therefore we should not commit the designation to memory in too final a fashion". Prescient words.

76mm GMC T70 as tested by Armored Board

Another view of Armored Board's T70

In the end, both engine and transmission were mounted on rollers to permit them to be serviced more readily. A cal. .50 machine gun was added on a ring mount for use against aircraft and ground targets. The individual torsion bar suspension utilized the same type wheels as in the T67, with a sturdier construction. The track was all-metal, single pin, rubber-bushed, having hardened surfaces at points of road contact. The power from the engine passed through the transfer case, propeller shaft, torqmatic transmission, controlled differential, final drives, and sprockets. The transmission was a manually selected, hydraulically operated, planetary gear train connected with a torque converter.

The first pilot vehicle was completed in the beginning of April 1943. By July 1943, all six pilots were completed and were being tested at Milford Proving Grounds, Camp Hood, Texas, and at Aberdeen Proving Ground. In general the performance was in accordance with the original specifications and the vehicle was successful in its field trials. The engine cooling, resistance to traction, braking characteristics, and fire control equipment were all found generally satisfactory. However, the weight had increased from the 33,400 pounds specified in the original characteristics to 39,600 pounds.

The principal deficiencies were the automotive type starter, short track life, and weak shock absorbers. A concentrated development of track components, particularly bushing sleeves and pins, increased track life to a satisfactory 2,000 miles under the most severe conditions. The shock absorbers were undersized for the weight of the vehicle, although they were the largest then commercially available. By mounting two shock absorbers on the front bogie, their life was considerably increased. A suitable shock absorber, designed especially for the vehicle, was added later. Experiments were made with track tension arrangements on sprockets and idlers to compensate for track slack and throwing when load pitching occurred.

Two views of 76mm GMC T70 Pilot #3, as seen at Aberdeen, 11th June 1943

Difficulty was encountered as the arrangement advocated by General Motors allowed no slack when dirt, rocks, or mud were picked up by the track and carried over the rear idler. Lack of sufficient improvement in track performance resulted in further studies being abandoned.

Because of the rubber shortage at the time, a brief study was made on the use of all steel bogie wheels. The very brief life offered by these wheels and the improvements in synthetic rubber made the change to all steel undesirable.

Nine rounds 76mm and cal. .50 ammunition stowed in front of the loader's position in the T70 as tested by the Armored Board

The 76mm gun and mount were generally satisfactory, with only a few minor faults. The gun and mount were given a formal proof firing test, functional firing test, jump firing test, and then subjected to a 1,300 round endurance firing test at Aberdeen. These tests were all satisfactory. It was found that Gun Mount T1 was extremely difficult to remove and to reassemble to the turret of the vehicle. The gun elevating mechanism vibrated loose from its mounting on four occasions, but this condition was corrected by increasing the size of the mounting bolts. The recoil guard covered the left recoil cylinder in such a manner that the filler plug could not be withdrawn. The balance plate at the rear of the recoil guard interfered with the normal loading of the gun, as the base of a round had to be raised above the plane of the bore to provide clearance for the loader's hand. The gun was moved two inches to the right of the turret center line to provide a more comfortable position for the gunner. The torque imposed on the light turret, when firing this high velocity gun, resulted in a "kick" in the hand traversing mechanism but this was not considered serious since power traverse was generally used. Deep water fording to a 6-foot depth was accomplished by the addition of fording stacks for the engine air intake and exhaust, and for the driver's hatch.

The vehicle was completely winterized with the quick heating type gasoline burning Petro heater for the engine compartment and transmission. A baffle was installed in the torque converter blower outlet to divert the warm air into the crew compartment. The ammunition stowage was unsatisfactory because the locking clips on the 76mm ready rounds were easily broken, the stowage compartments in the turret for cal. .50 ammunition were too small to allow easy stowage and withdrawal, and the 76mm ammunition stowed in the sponsons was not adequately protected from small arms fire.

The periscopes in the driver's and assistant driver's hatch doors struck the roof when the doors were open and the seals on the two doors were not satisfactory, because the rubber seals pulled loose from the retainers. The pintle of the vehicle, a standard ¾-ton, 4x4, truck pintle, was too light for practical use. Also, the pintle was rigidly mounted instead of being free to swivel.

One of the difficulties in the development of this vehicle arose because at the initiation of the project a quota of 1,000 vehicles had been set for completion by the end of 1943. Production vehicles started corning off the line in June 1943. Testing of the pilots and elimination and correction of deficiencies had to be greatly expedited in order that the changes could be incorporated at an early date in the production vehicles.

However, in January 1944, when the development phase of this project was concluded, the quota of 1,000 vehicles had been filled. In July 1943, the Ordnance Committee had issued a directive for test of 76mm Gun Motor Carriage T70 by the Armored Board, Fort Knox, Ky. The test was to determine the capabilities and limitations of the vehicle with a view to its possible use as a light tank. In January 1944, the Board reported that the vehicle was not suitable for use as a light tank because it had insufficient armour protection (.30AP would penetrate the turret at under 75 yards), it lacked turret top protection and antipersonnel weapons, and had no protection for the crew while firing the machine gun. The tests conducted by the Armored Board were limited, since the Tank Destroyer Board was conducting more extensive tests. The Armored Board stated that any correction of the deficiencies, which then made the vehicle useless as a light tank, would add materially to the vehicle weight and would reduce the chances of satisfactory performance. They did comment favourably, however, on the low silhouette, the compensation of the idlers, the torsion bar suspension, and the engine/transmission mountings which facilitated maintenance.

In February 1944, the Ordnance Committee recommended standardization of the T70 as 76mm Gun Motor Carriage M18. It also established that vehicles with serial numbers of 658 to 1095 be modified to incorporate a redesigned lower differential gear ratio before overseas shipment. This was a mandatory change which increased the vehicle's mobility in all conditions before the vehicle be distributed to the troops. An additional 35 changes were recommended varying from more trivial issues such as the tray for the cal. .50 being designed for 50 rounds whilst the stowage was for 100-round boxes, to more important safety features such as a gunner's shoulder guard or an impulse firing relay modification to prevent slamfires (Unintentional firing of the gun caused by closing the breech). As many of these features as possible were to be incorporated in those vehicles. All vehicles below serial number 658 were to be returned to the factory for modification.

The recommendation for standardization and disposal of vehicles was approved in March 1944, by the Ordnance Committee.

Ballistic tests by the Armored Board showed that at 75 yards (top), about one round in three of cal. .30 AP would penetrate the turret of T70, and at 50 yards (above) the majority would

141

The Tank Destroyer Board, Camp Hood, Tx, had conducted tests of four late production models of the T70 in order to determine the adequacy of the mechanical corrections made on these models, as recommended after tests of pilot vehicles. The Board also was to determine the comparative tactical performance between vehicles having greater low-speed drawbar pull as a result of gear reduction in the differential.

The test of these four production models indicated that the modifications made to correct deficiencies in the earlier models were successful, but the failures in this test showed that further corrections were required, particularly as to drive sprockets and transmissions. Failures of the transmission brought to light two inherent defects: The need of a positive lock washer for the rotor shaft nut and a transmission transfer case-driven gear bearing capable of standing up under 4,000 miles of operation. The drive sprocket needed further strengthening at the teeth by thickening, reinforcement, or change in flame hardening. Shock absorber failures were not disabling, but threw stresses in other parts and constituted a resupply problem. These shock absorbers were more rugged than the original type, but required further correction.

This page: Three views of T70 Pilot #35, as seen 23 August 1943 in Aberdeen

The Tank Destroyer Board recommended that a more rugged transfer case-driven gear bearing be developed, that the drive sprocket be strengthened, that the shock absorber be made more durable, and that a dependable lock washer for the transmission rotor nut shaft be made.

The Tank Destroyer Board also conducted tests of the performance of the M18 with added weight, to determine the capacity of the vehicle to carry additional armour without modification to suspension and power train. A late production model was operated with various weight additions. The Board, in August 1944, concluded that supplementary selective armour weighing up to 2,000 pounds would not adversely affect tactical mobility. The Board recommended that a study be made of the application of supplementary armour in selected locations on the vehicle, applied as spaced armour, at an angle to the basic armour, to utilize the maximum deflecting properties of added plate. This recommendation was concurred in by the Tank Destroyer Center Headquarters, but no further action was taken.

Microphone Switch   Observer   Extension Cord
SW-141 and Headset
Jack JK-26

Top: The full issue of equipment for M18. Above left: T70 showing bustle stowage. Above right: Photograph from Tank Destroyer Board firing tests, Camp Hood, Tx, showing the gun commander dismounted from the vehicle in order to conduct observed fire. This would mitigate the obscuration effect from the main gun blast.

In spite of minor needed alterations to 76mm Gun Motor Carriage M18, the vehicle as standardized was considered very satisfactory, the general performance was excellent, and the riding qualities in particular were superior to those obtained in any previous tank destroyer. The outstanding features of this vehicle, such as the torsion bar suspension, torqmatic transmission, engine and transmission mountings to allow quick removal, large air cleaners, and bulged turret for balance, were recommended for study for possible inclusion in future vehicles. The use of the torqmatic transmission was to be studied in conjunction with the use of the aircraft type radial engines, since it greatly extended the useful life of this type engine in tanks. The loading on the engine with the torque converter transmission was very similar to that obtained in the aircraft installation for which the engine was originally designed.

The issue of the ammunition protection was again looked at in September of 1943. Given the thin nature of M18's armor, it was theorised that small calibre projectiles such as .50 or .30 might penetrate through to the ammunition stowage. As a result, firing tests were held to determine the effectiveness of water-jacketed stowage (commonly known as 'wet stowage'). However, the holes created by the small projectiles resulted in such a low rate of water flowing into the ammunition that it had no particular effect on preventing burning.

The vehicle's introduction into service, however, was not without issue. When the first T70s arrived in Italy, the big concern was the fact that it looked rather a lot like a Panther, with its long gun, somewhat angular shape, and large independently sprung road wheels. Fratricide concerns were up front.

Further, the troops simply didn't want the thing. It had a lot less armour, and they trusted the M10. An internal memorandum from COL Montgomery, president of the Tank Destroyer Board, to MG Bruce dated mid-August 1944 lamented the introduction into service of the vehicle. And, more importantly, the fact that the M18 was about to be cancelled as, in the words of COL Dean over in Army Service Forces, "If the troops won't use them, it is useless to produce them."

The M18, designed to fit a new concept [...] should have received its battle try-out under favorable auspices. That would mean a dozen battalions, equipped with the latest and best models of the weapon, manned by at least average personnel, well trained <u>on this weapon</u>, under commanders familiar with approved tactical doctrine, adequately worked out as was done here in the Center, School and Board.

The above has not taken place. The first weapons went over "in bulk" and were given to a towed unit trained on half-tracks. We have never heard whether they got into the fight or the results if they did. [...]

There seems no reason, if the thing had been properly handled, why a few weeks later a few M18 TD Battalions should not have gone into the fighting. Instead, we hear that the "troops don't want them". Units are apparently being canvassed as to whether they want this new weapon in preference to the ones they now have.

Theoretically, preparation for reception of the M18s in the active theatres was made. It consisted of arrangements for sending civilian representatives from the Buick organization into the theaters. So far as I can learn, no officers of special M18 training have been sent to staffs in England or Italy. The troops have been left to find out the capabilities and limitations of this vehicle by trial and error.

To straighten this out, to save the Government's investment, and to get this outstanding new weapon into the fight in the hands of personnel <u>not hostile to it</u>, and trained in its proper use and mechanical handing, the matter should, I believe, be brought urgently to the attention of AGF. [...] A plan should be made to have one or more "missionaries" sent to the staffs of theater commanders. (This should have been done months ago by AGF without prompting)
[...]

I recommend that you present the bungled handling of the entry into battle of the only tank destroyer not an expedient, and ask that something be done to pull the fat out of the fire even at this late date[...]

Of course, in the end, M18 did achieve acceptance, though ultimately not universally so. M10 and its thicker hide, and M36 with its bigger gun both had their advocates over the nimble, fast-hitting Hellcat.

Accessories

On 25 May 1943 the Chief of the Development Division directed that a study be initiated on the possibility of floating the 76mm Gun Motor Carriage, T70. "To provide flotation equipment which will permit the 76-mm Gun Motor Carriage T70 to negotiate, under its own power, deep rivers and expanses of ocean." This study was made and tests were conducted to determine the feasibility of propelling the vehicle by means of the tracks. As a result of these tests, it was discovered that approximately 1,500 lbs of drawbar pull could be obtained by rotating the submerged tracks with the upper portion skirted.

The first test of a T70 with floats at Ford's River Rouge Plant. Unlike the production T7, these early floats were single-piece, non-folding, and steering was by tracks only

On 29 November 1943 a request was made to Buick Motor Division to construct a set of floats to fit the front and rear of the vehicle. 29 December 1943, a test of the floating unit in the boat slip at the Ford Motor Company, River Rouge plant resulted in a speed of 4.24 mph and a turning radius of 74 ft.

1 January 1944 was the starting date for the design and construction of four additional pilot sets of flotation equipment, and on 15 February 1944, the first pilot incorporating a mechanical float release was demonstrated at Ft. Story, Va. to a group of Army and Navy personnel including representatives from the different theatres of operations.

Two M18s with T7 Swimming Devices (also known as Ritchie or Berg Devices) seen at the Amphibious Warfare Board test site in Fort Ord, CA (Top) and Aberdeen Proving Ground (Bottom). Not in the top photo how the already low rear clearance is exacerbated by sinking into sand.

In February 1944, when Swimming Device T7, used on 76mm Gun Motor Carriage M18 was classified as a limited procurement type, it was directed that the adaption of a gun stabilizer be investigated to improve accuracy of fire of the 76mm gun during water operations. An installation of the stabilizer was made and tested both at Fort Story, Va., and at Aberdeen Proving Ground. A vehicle equipped with a gun stabilizer was then sent to Camp Hood, Texas, for service test by the Tank Destroyer Board.

Top: M18 with T7 Swiming Device at Fort Ord. Above: Two views of M18 with T7 under test. In the background on the right can be seen what appears to be an M4A1 with a T6 Swimming Device

Subsequently, on 16 March 1944 a decision was reached to build 250 of these kits as soon as possible. More suitable rudders, splash shields, gun blast reinforcements, fairing, float folding means, fording stacks, running lights, and float releases were then developed and incorporated for test on the Numbers 2, 3, and 4 pilots. Gyro stabilizers for the 76mm Gun were installed in the Numbers 3 and 4 pilots, tests of which showed conclusively that gun stabilization was necessary for accurate firing from the water.

Insufficient freeboard was provided for operation in high surf, and the open top turret and torque converter outlet were not suitable for operation in high seas. A modification was trialled which reinforced the pontoons to reduce damage from the blast from the main gun, they were filled with foam, and a canvas cover for the turret roof was developed. Also added were LVT-style grousers. Most of the modifications were considered to be an improvement over the original device, but the grousers were too weak to merit the extra 0.5mph in water. Drawings of the subject device were released for production to the Engineering & Manufacturing Division on 22 March 1944. Device T7 was reclassified from "Secret" to simply "Confidential" 20 July 1944.

Army Ground Forces approved the Tank Destroyer Board 's recommendation that 76mm Gun Motor Carriages M18 fitted with Device T7 be equipped with a gun stabilizer before shipment to theatres of war. However, since the ultimate use of each vehicle could not be ascertained at the time of processing through the Ordnance Depots, the gyrostabilizer was not to be installed in these vehicles at the depots, but was to be included in the Device T7 kits for field installation. A directive to this effect was issued in October 1944.

An M18 underway. This appears to be undertaking firing tests. Turret splashguards do not appear to be fitted.

In an attempt to give the commander some protection when firing the machinegun, a clamshell covering was developed and tested in November 1950 as part of a broader program to protect .50 cal positions in the Army's armoured fleet as a whole.

A protective dome hinged at the rear of the ring mount covered the operator's head and gun controls when the machine gun (.50 caliber) was in the ground-firing position. In the antiaircraft position the protective dome was opened, extensions to the machine gun handle permitting control below the ring level. Traverse movement of the ring was controlled by the operator's shoulders. A pintle support on the cradle support allowed ten degrees of secondary traverse.

This page: Three views of the clamshell covering for the .50 cal position. Above left and left, closed in ground firing position. Below right, the clamshell open, for the anti-air role.

The turret gave the gunner protection while firing in the ground-firing position only. The change from ground to anti-aircraft position was found to be awkward and expose the gunner, with his having to push back the protective dome, reach out and use considerable force to move the gun up into the antiaircraft position thereby exposing himself to enemy fire. Visibility was poor and the gunner was very crowded. In the end, the idea as a whole was abandoned.

On the matter of overhead cover as a whole, the experiments by the Tank Destroyer Board started in 1944 applied to the M18 as well. One of the first concerns was that of cooling: Given the air flow through the vehicle, would closing off the top result in issue? To experiment, an M18 fitted with a canvas roof in October 1944, it was noted that in water the temperatures in the cylinder heads rose almost 20 degrees Fahrenheit. It was also noted that the M18 was not designed to withstand overhead fire regardless, as one look at the engine deck would confirm.

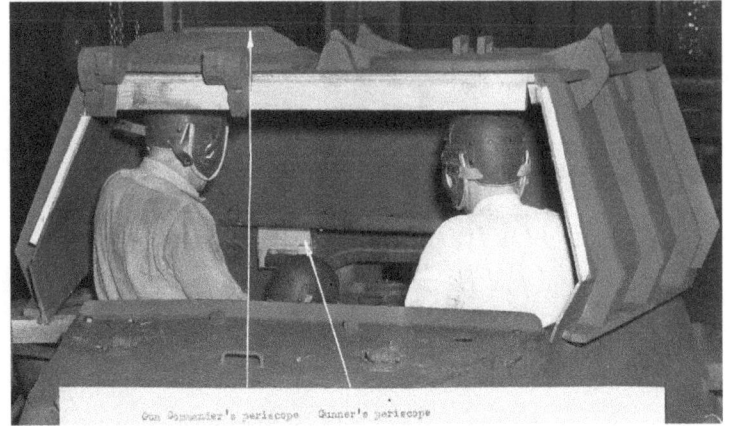

Above: Some of the earlier turret enclosure designs for M18 were, perhaps, wrong turns
Below: The first turret roof trialled at Camp Hood. Bottom: The first roof next to an unmodified M18.

Above: Two more views of the first turret cover

Closed-in sides
MG ring cover
MG ring cover supports

Above: Four pictures of the second roof trialled. In all but the top left, the removable .50 cal. protection has been added. The ring cover would be stowed on the hull rear.

After some aborted designs which permitted full standing height in all positions, the first turret roof trialled consisted of simple hinged sheet metal, placed with small vision slits around the sides. This was soon discarded in favour of a higher roof with up to a 10" visibility gap near the centerline. A further development added a removable "Chinaman's hat" machinegun cover, and some armour protection around the sides of the commander.

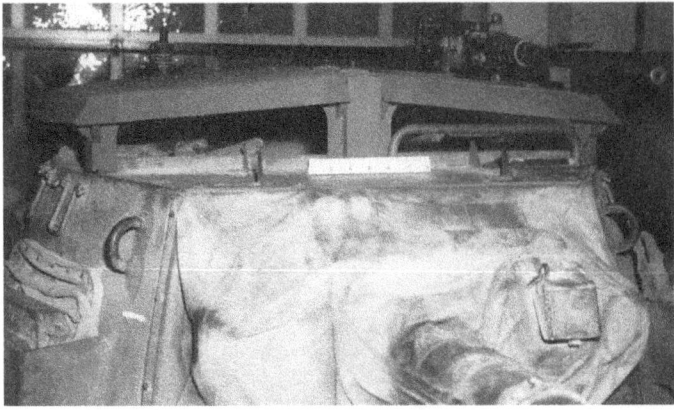

Above: Front and rear views of the second roof trialled, both without the additional .50 cal. protection

In December of 1944, an M18 was fitted with a Target Range Timer T1. This device linked into the interphone system and the firing circuit. The purpose of the device was to aid the gunner in estimating lead on a moving target. To operate the system, a crewman would first dial into the machine the range to target, from 500 to 3,200 yards. The gunner would then lay onto the moving vehicle, or just in front of it, and stop turret movement. When the vehicle drove into the reticle, the gunner would then either depress the firing solenoid or a crewman would manually trip the timer. The timer electronically then counted down the time of flight of an M62 APHE round, and at the appropriate moment that the round would have reached that range, a tone sounded over the intercom system. The gunner could then observe how far in front of the reticle the target had moved, re-lay on the target with that amount of lead, and then engage. As an expedient test, the timer was mounted on the M18 wherever it would fit. An initial attempt to mount it on the .50 cal mount resulted in the vibrations affecting the range dial, but a later mounting on the turret bustle proved more acceptable. Inside the bustle next to the radio was considered the best theoretical location, on the off-chance that anyone decided that the thing was worth including. It would appear that in the end, nobody did.

The Target Range Timer T1 as seen in its original position (left), and then on the bustle (right). In the background of the right photograph can be seen what appears to be the hull of a Panther.

In February 1945, a study was launched by OCM 26715 to ascertain the problems involved and the equipment required for the disassembly and stowing of the GMC M18 and Light Tank M24 in the C82 Packet transport aircraft for airborne shipment. This proved to be a very short study after an individual with a measuring tape determined that the M18 was too wide to fit into a C82. The project continued for a while longer using the M24 light tank.

In July of 1945, a project was started for both the Light Tank M24 and GMC M18 to try to reduce the ground pressure of the vehicle to some 7psi. Initial tests conducted the second half of 1945 were of a 21" track named T82. Durability was a problem, though, so a new type of track was created, T86. The difference in weight for the standard T69 track was the difference between 3,254lbs/pair for the standard, and 5,369lbs/pair for the T86.

A few changes were required in order to get the track to work. Firstly, the fenders had to be modified to allow the clearance, secondly the sprocket wheel needed replacing, and finally, it was discovered that the centre guides were too high to clear the hubs of the roadwheels. The first was a simple fix, new sprockets had to be made, the roadwheels were replaced by those of M24 (about an inch greater in diameter) and the idler adjustment system needed to be swapped between sides and reversed.

Testing was conducted firstly in Aberdeen Proving Ground, and then by the Armored Board, since the Tank Destroyer Board had been dissolved. The conclusion was that the wider tracks and added weight slightly increased the rolling resistance of the vehicle, and reduced its hill-climbing ability in particular. The project was terminated.

A concern of the design of M18 was that the engine exhaust might be being drawn back into the engine intake, and thus affecting cooling. Anti-recirculation exhaust louvres were fitted to a vehicle to try to reduce the effect. In the end, it was concluded that there was no practical benefit, as the temperature of the engine only reduced the average cylinder head and base temperatures of the engine by about 5°F

Top: Modified M18 with 21" track
Right: The anti-recirculation louvres, 02 Oct 1943

An M18 in Camp Hood demonstrates the rails used to remove the transmission.

The utility vehicle variant of the M18 probably deserves mention in this volume as the idea and testing for it was the product of the Tank Destroyer Board. On 25th Feb 1944, the Board submitted to Army Ground Forces a report on alternate uses of the T70 chassis.

In the testing of the 76mm Gun Motor Carriage T70 it was constantly demonstrated that none of the other vehicles in tank destroyer units were capable in tactical operations of keeping up with the T70 and aiding in the exploitation of its full tactical capabilities. The T70 frequently over-ran its reconnaissance and security and had to be held up. Commanders could not properly perform their functions using the transportation provided.

Since the tests included the best available wheeled and multiple-wheeled vehicles (i.e. ¼-ton truck, the six-wheeled M8 and M20 as well as the half-track (M3) and the full-track Universal Carrier, T16, and all were found inadequate, the T70 chassis was made a subject of experiment for a possible solution.

The turret, turret ring, and gun mounting were lifted off, leaving the deck in place. The auxiliary generator with its gas tank was eliminated as no longer needed (no power traversing). The sub-floor was taken out and storage batteries and fixed fire extinguishers placed elsewhere. A tunnel was put in over the propeller shaft and other piping.

Adaptations of the resulting stripped vehicle were then experimentally made to produce in turn:
    a. An armored command-reconnaissance vehicle.
    b. An armored personnel carrier (Security)
    c. An armored ammunition cargo carrier
    d. An armored ammunition prime mover for artillery towed guns
    e. An armored general purpose carrier, having approximately 80 cu. ft. of space available and capable of carrying 8 tons payload.
    f. An armoured mount for an anti-aircraft machinegun turret.

The differences between "a" and "b" were that the seating in "a" was more modular, and could leave room for up to 90 rounds of 76mm when carrying only four people. The armoured ammunition carrier would remove the side seats, leaving a cargo capacity of up to 297 rounds of ammunition. The artillery prime mover version proved in effect to also be the first variant, carrying both gun crew and ammunition. They never did end up making the anti-aircraft version.

A Tank Destroyer platoon incorporating an M18 CRV under test in the command role, 24 February 1944

In a test over a 2,100 yard course, which included swampy, low ground, 600 yards of ploughed field, steep rocky hillsides and thick growth of scrub pines, the times for the various vehicles were as follows:

T-70 C-R Variant: 4'20"
T-70 Destroyers: 7'10"
¼-ton truck, M20, Half-Track, 1½- ton truck… all 9' 10".

Observers were reported to have been impressed by the C-R vehicle's low silhouette and the inability to distinguish the purpose of the vehicle.

A proposed platoon organization would have seen the Platoon commander's ¼-ton tuck, and the Dodge 1½ton 6x6 carrying the ten man security detail and towing a trailer carrying 52 rounds of ammunition replaced by two T70s. The Board favoured this organization, noting that the identical chassis would greatly simplify maintenance operations, and recommended that such turretless T70s replace all wheeled vehicles which would ordinarily accompany 76mm Gun Motor Carriage T70s into combat.

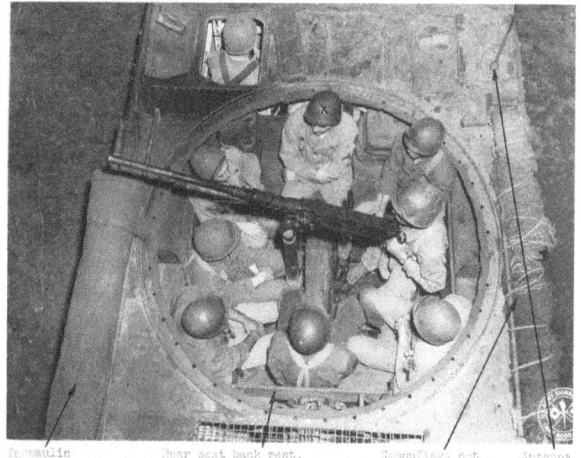

Left: The M18CRV climbs a muddy slope on which M20 failed. Note the canvas covering the crew. Right: Proposed seating of an eight-man security team. In this configuration, the batteries are in the assistant driver's position. An alternate proposal would see the batteries placed where the soldier marked with the "X" is seated.

The correlation of the development of the M39 Armored Utility Vehicle can likely be inferred from this first initiative from Camp Hood.

Two views of the M39 AUV for comparison.

A final note on the name "Hellcat." Although this originally appears to have been a product of Buick's marketing machine, the name stuck, gaining acceptance, at least as far as General Barnes was concerned, by November 1944 in his order assigning official nicknames to vehicles.

76mm Amphibian Motor Carriage T86/T86E1

T86E1 Right Side View

In December 1943 and January 1944, meetings were held on the Ritchie Project, attended by representatives of General Staff, Army Service Forces, Corps of Engineers, and Navy Department, in which the development of various amphibious devices were authorized. As a phase of these activities, the development of three pilots of an amphibious gun motor carriage based on the 76mm Gun Motor Carriage M18. It was desired to develop an amphibious vehicle having high firepower with good performance on both land and water without the necessity of any additional equipment or preparation. The use of the M8 HMC turret on the LVT hull was a start, but not ideal.

Marmon-Herrington company, of Indianapolis, was given a contract on 26 January 1944 to build three pilot vehicles based on M18. Naval architects Sparkman & Stephens Incorporated acted as consultant on the naval portion of the vehicles. The work consisted of removing the hull plate of the 76mm GMC M18 down to the sponson line and substituting a larger amphibious type hull of lighter construction, the final drive gears changed to give approximately 14% more reduction, and the suspension modified slightly to accommodate a new 21" wide track, skeletonised from the M24 light tank.

In February 1944, the Ordnance Committee recommended the initiation of a project for the development of necessary equipment to provide for the floating of combat and transport vehicles across bodies of deep water. As a phase of this development, the Ordnance Department began work on a 76mm Amphibian Gun Motor Carriage M18. The upper hull structure above the sponson line and all non-essential heavy parts were eliminated, and a light weight, amphibious type hull was fabricated on the remaining chassis, equipped with the turret used on the M18 and a 76mm gun with a gyrostabilizer. The procurement of two pilots was formally approved by the Ordnance Committee in September 1944. At this time it was also recommended that a vehicle incorporating the best features of the two vehicles, and mounting a 105mm howitzer, would be developed after test of the T86 and T86E1.

The first pilot vehicle mounting the 76mm gun was designated 76mm Amphibious Gun Motor Carriage T86. This vehicle was propelled in water by means of the tracks with proper skirting of the track tunnel. The second pilot, designated as T86E1, was to be similar to the T86 except that it was to be equipped with twin 26" diameter screw propellers driven from the rear transfer case. Both of these pilots were equipped with the 76mm gun and turret, as used on the M18. A third pilot, designated as T87, was to be equipped with the 105mm Howitzer mounted in the same turret. The building of this pilot was held in reserve until it was determined which chassis, T86 or T86E1, was the most satisfactory.

The first pilot T86 was completed for initial tests on 15 June 1944, and was demonstrated at Aberdeen Proving Ground on 20-23 June 1944. The land performance was about the same as the Gun Motor Carriage M18. In water it floated with about 15" of freeboard and the water operation was satisfactory. After several modifications were made to improve the vision, water steering, bilge pump installation, etc., the vehicle was shipped out to the Landing Vehicle Board, Fort Ord, California, to obtain additional test data. The Landing Vehicle Board reported that the vehicle was sea worthy; however, that it was not suitable for military use in its then-present form.

The Landing Vehicle Board didn't like T86 at all:

Due to the size of the rudders and the short arm by which the rudders are actuated, extreme force was required to turn the rudders.

Continued water-borne operations indicated that vehicle is sea worthy and does not tend to be top-heavy. Steering difficulties were the rule rather than the exception. The reaction of the vehicle to rudder steering alone was completely unsatisfactory. The lag between the application of steerlng effort and the response of the vehicle was so marked that as a result of over-control the vehicle proceeded along its course by a series of short legs at right angles to each other. Even an experlenced driver could hold a course no closer than 30 degrees. [...] Further experience indicated that best control of vehicle while waterborne is obtained by use of track steering alone. In a fairly calm sea an experienced driver can hold a course within 5 degrees

This page: Three views of 76mm AMC T86 at Aberdeen

Repeated landings through and launching into the surf showed the vehicle capable of operation in beakers up to ten feet high. When making landings and launchings, the rudders were worse than useless. When entering the surf, the bow of the vehicle is often lifted by an incoming breaker as the vehicle is almost water-borne - This throws the stern down and in most cases bends one or both rudders when they strike the sand. In landing, the rudder is of no value, as the relative motion of the vehicle and the water is reversed - the water is traveling in the same directlon as the vehicle and at a greater speed. The transmission with which the T-86 is equipped is well adapted for for landing through the surf. The practically instantaneous shift and ease of making the shift at the critical instant when the tracks contact the sand give positive control without possibillty of stalling the engine due to poor clutch and throttle technique.

Two views of T86 under test in Fort Ord, 13th February 1945

156

During firing, it was established that the 76mm gun could be fired broadside with no adverse effects on the crew or vehicle. When firing over the bow, muzzle blast so deflected the foredeck that the driver in the vision cupola received a solid thump on the head from the deck above him. The gyrostabilizer worked satisfactorily.

On the other hand, the vehicle was found to be very maintenance-intensive. "The chief faults which appeared were generally the result of lack of appreciation of the true character of amphibian operation." Internal stowage was unworkable. Top speed on land was limited to 20mph due to pitching of the vehicle. Attempts at increasing the speed by adding grousers to the track met with limited success.

Above: T86 in cleaner days. Below, comparative views of T86 (left) and T86E1 after modification to single-screw (right)

The T86E1 pilot was completed in November 1944. The preliminary tests on the vehicle performed at Indianapolis were disappointing in that only a slight increase in water speed was obtained with the twin screw propellers over that obtained with the tracks on the T86. This was thought attributable to the steep propeller tunnel made necessary by the contour of the Gun Motor Carriage M18 chassis. In view of this it was decided to raise the engine to allow a more efficient tunnel and to incorporate a single large screw propeller to simplify the drive. A number of additional improvements were made based on the experience of the T86. This modification work was completed in July 1945, at which time it was shipped to Aberdeen Proving Ground for engineering test. A substantial improvement in performance was obtained by the modification to the propeller installation. However, the maximum speed was still about one and a half miles per hour short of the seven and one-half mile per hour objective. This was believed to be due to the very high hull resistance and the difficulty in getting sufficient water to the propeller due to the protection required.

Active work on the T87 was withheld until November 1944 when the results of the tests on the T86 were known. In view of the relative success on the T86 at Aberdeen, it was at the time tentatively decided to use track propulsion for water operation of the T87. A number of improvements were incorporated and the vehicle completed early July 1945.

Shipping instructions were issued on 26 September 1945 to ship the T86E1 to the Landing Vehicle Board, Fort Ord, Calif. for service board tests. In January 1946 the T86E1 shipping orders were cancelled and the vehicle was withdrawn from the Service Board Test. This action was based on the recommendation of Aberdeen Proving Ground, that no further tests be conducted on this vehicle due to its very unsatisfactory performance when operating in water.

As determined from engineering tests at Aberdeen Proving Ground, the steering and propelling arrangements of the 76mm Gun Amphibian Motor Carriage, T86E1 were not suitable to enable controlling the vehicle when afloat. To correct this deficiency would require a redesign of the vehicle; however due to reduction of development funds, it was not feasible to continue further development. As a result of the foregoing, the proposed test of the T86E1 by the Army Ground Forces Board No. 2 (Landing Vehicle Board) at Fort Ord , California was cancelled and the vehicle was placed in the museum at Aberdeen Proving Ground for historical purposes.

This page: Multiple views of T86E1. Lower right, a close-up of the screw tunnel, with cracks highlighted by circles

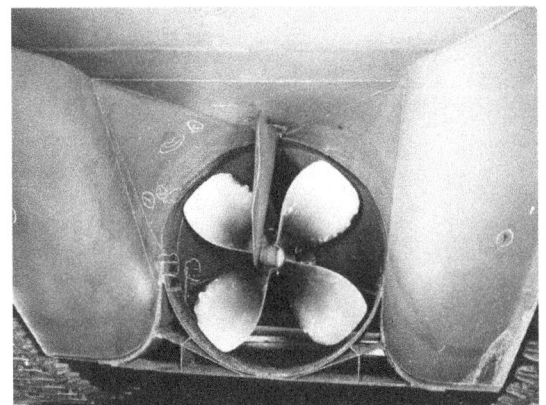

76mm Gun Motor Carriage T72

The object of this project was to develop a turret and gun mount for the 76mm gun to be mounted on the chassis of the Gun Motor Carriage, MIO. The development project was initiated as a result of the following conditions which existed in the production 3" GMC, MIO.

a. The turret and gun mount with no counterweight was approximately 190,000 inch pounds out of balance. This unbalance made it impossible to traverse the turret satisfactorily on slideslopes and necessitated the addition of a 3,600 pound counterweight to the rear of the turret in order to reduce the unbalance.

b. The relationship between the traversing handwheel, the elevating handwheel, and the direct sight telescope was unsatisfactory, making it very difficult for the gunner to properly lay the turret.

c. Due to the design of the turret and the width of the 3" gun mount, the crew space in the fighting compartment was very cramped.

At the very end of December 1942, the initial request was sent from the Components Section of the Arms and Armament Unit of Ordnance Branch to request that a turret from 76mm Gun Medium Tank T20 be modified as follows:

Demonstrating the turret crew positions.

    a. Trunnion support side plates, 2" thick. Distance between outside of these plates to remain the same, ring to be welded on side plates to keep 3" bearing surface for trunnion pin.

    b. Side plates for turret body, 1⅛" thick.

    c. Rear body of turret plate, 1⅛" thick.

    d. No pistol port.

    e. Front sloping top section, ¾" thick.

    f. No center top section.

    g. Top section over radio bulge, ½" thick.

    h. Use same turret base ring as Medium Tank, T20, Welded turret.

    i. Thickness of shield, 2" where possible.

    j. No co-axial machine gun.

    k. No power traverse. New process gear box to be used temporarily for manual traverse.

    l. No radio in bulge. Provide space for stowage of 76mm ammunition in turret bulge. Use basket type ammunition container developed for 105mm ammunition.

    m. For present redesign, use Medium Tank, T20 direct sight telescope.

    n. Details of the gunner's, loader's, and commander's seats to be supplied at a later date.

In common with the medium tank M4E6 program, which also was to use the medium tank T20 turret, the donor vehicle was changed to Medium Tank, T23. As a result, the turret and gun mount of T23 was revised by the Tank-Automotive Center, Development Branch, as follows:

a. The thickness of the turret sides and rear section was reduced from 2½" to 1⅛"

b. The thickness of the gun shield was reduced from 3½" to 1½".

c. The turret top was omitted.

d. A bulge was added to the rear of the turret which acted as a counterweight and also provided room for the stowage of 27 rounds of 76mm ammunition.

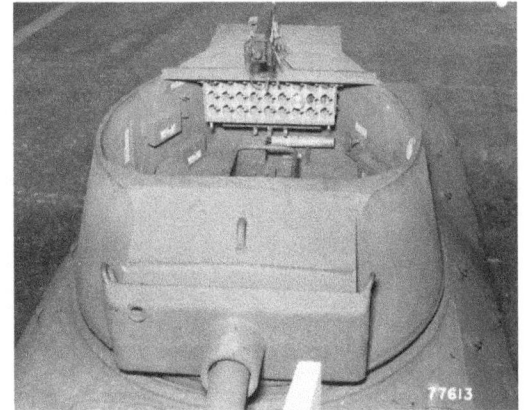

Left: 3/4 view of the T72. Above, a view of the turret rear absent the clutter of Ford engineers.

After the drawings of the turret and gun mount for the 76mm GMC T72 were completed, these drawings were delivered to the Ford Motor Company and two pilot turrets and gun mounts were manufactured. Upon completion of the manufacture, the turrets and gun mounts were installed on the standard 3" Gun Motor Carriage, M10A1, chassis and the first vehicle was shipped to Aberdeen Proving Ground on 19 March 1943, the second on 20 April 1943. Somewhat bizarrely, the OCMs officially approving the T72 project and authorizing the procurement of the two pilots were 19953 and 20146, dated 22 Feb and 03 April 1943 respectively. A review of the various documents in the archives indicate that a number of actions related to the development of new equipment were carried out on the basis of nudge-and-wink and/or verbal approval, if not simply under the 'better to beg forgiveness than ask permission' philosophy.

More T72

As a result of proving ground tests, which were completed by mid-July 1943, it was determined that the T72 was a satisfactory vehicle and that it held the following advantages over the M10:

a.   The turret and gun mount assembly was almost perfectly balanced with a resulting weight saving of approximately 4,350 pounds as compared to the turret and gun mount weight of M10.
b.   It was possible to obtain a very satisfactory location of the traversing handwheel, the elevating handwheel and the direct sight telescope. By use of the standard Medium Tank, M4 gear box, a very satisfactory traversing mechanism was also obtained.
c.   Due to the difference in size of the ammunition and by using the turret bulge for storing 27 rounds, it was possible to obtain an increase of 46 rounds of ammunition in T72 as compared to the 3" Gun Motor Carriage, M10.
d.   Due to the fact that the 76mm gun mount was 7" narrower than the 3" gun mount and by bulging the sides of the turret body instead of bringing these sides back straight, it was possible to obtain considerably more room for the crew in the fighting compartment.

The only noted disadvantage compared to M10 was the fact that the gunshield afforded somewhat less protection, the M10's being some 2½" compared to T72's 1½".

Gunner's position, T72. Note the lack of as much as a footrest.

It was decided not to release the 76mm Gun Motor Carriage, T72, to production, nor to send this vehicle to Camp Hood, for Service Board Tests for the following reasons:

1. Vehicles of the Gun Motor Carriage, M10, type were to be replaced in production by the 76mm Gun Motor Carriage,T70.

2. Due to the necessity of retooling for the production of the turret and gun mount of the Gun Motor Carriage, T72, it was decided that if a 76mm installation was desired on the Gun Motor Carriage, M10, chassis, to use the turret and gun mount then being developed for the Gun Motor Carriage, T70.

Although this ended the T72's development program, the vehicles were still put to some use. One year to the day after the first request to modify a T20 turret for T72 was made, a new program was started to investigate the possibilities of non-recoiling mounts which could eliminate the necessity for a recoil mechanism, thus creating more room in the fighting compartment for the crew as well as lighter weight.

The theory behind the idea was to attach the gun to the vehicle so that the mass of the complete vehicle would absorb the recoil energy instead of using a recoil mechanism for the purpose. T72 was used because "it was available and not being used for any other purpose".

Above: Another Ford engineer is visible in gunner's position, head to the sight, and hands on the controls.

The design of this gun mount consisted of modifying a standard 76mm gun breech and designing a cradle which fastened the gun securely to the turret, thus transmitting the recoil energy to the vehicle and thereby using the complete vehicle as a recoiling mass. The mount was designed to take both the 75 and 76mm guns, but since the 75mm gun was declared obsolete, work was concentrated on the 76mm gun installation. The cradle was manufactured at Rock Island Arsenal and the breech ring was manufactured at Watervliet Arsenal. Initially the program was for a recoilless gun mount, but due to confusion, presumably with what we commonly know today as recoilless rifles, a nomenclature change was made to become "non-recoil gun."

The designs were completed by April 1944, and the mount, fitted for the 75mm gun M3, was shipped to Aberdeen where it was mated with the T72. Successful testing of this mount saw on 12 August 1944 the decision to move ahead with the 76mm gun variant, and adding to the design a complete fighting compartment, OCM 25393. This would be Mount, Gun, Combination, T116.

The design was started on the complete fighting compartment in OCO-Drafting Room but higher priority projects utilised most of the manpower and thus the development was slow. Upon the ending of hostilities it was believed that there as no immediate requirement for this type of gun mount and in view of the fact that steps were being taken to reduce the number of development projects an OCM was written on 20 August 1945 to terminate the project.

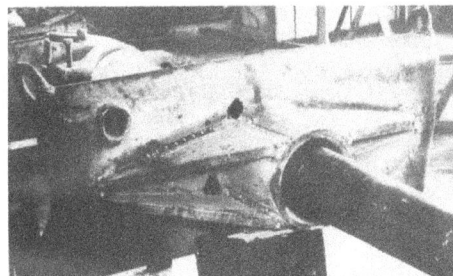

Above: Two views of the 75mm non-recoil gun mounted in T72.
Right: Two views of the uninstalled gun and mount.

90mm

90mm Gun Motor Carriage T53/T53E1

90mm GMC T53 with gun in travel position, 22 October 1942

In January 1942, a recommendation was submitted to develop a self-propelled mount for the 90mm gun in an anti-tank role, in response to reports from North Africa about the effectiveness of such a mount for the 88mm that the Germans were using (That this vehicle was the Flak 18 mounted on the 12-ton half-track was not identified until June). This was shortly approved, and steps started being taken to mount the gun on a medium tank chassis. Although it was considered to put the 90mm into the turret of M4, in order to put this gun in the turret, the base ring, gun mount, and recoil mechanism would have to be completely redesigned as no existing components could be used.

By June 1942, a contract was drawn up with Chrysler for produce a pilot model of a 90mm antiaircraft gun on the medium tank chassis. The mount was to destroy none of the major characteristics of the 90mm gun. It would have 80 degrees elevation, minus 5 degrees depression, 360 degrees traverse, and carry 100 rounds of ammunition. Speed was to be 30 miles an hour with a Wright 400 h.p. engine.

Authority was given 3 July 1942 for the diversion of two Medium Tanks, M4 and two 90mm guns and carriage, M1 and M1A1. Some guidance was helpfully provided: "Look into: Changing the air cleaners to make room for at least 50 gals. of gasoline; Eliminate the assistant driver's seat and place two batteries in this location; Make the gun base an integral part of the tank; Redesign the rear end so that the 90mm gun can be used as an anti-aircraft weapon; Use a pintle for towing ammunition and fuel trailers; Heaviest armor plate to be ⅞" thick top and bottom to be ¼" thick; Wright R-975 Engine to be used." The second pilot, a production model, would be begun only when the first was found to be satisfactory.

OCM 18495 of 16 July 1942 recommended development of 90mm Gun Motor Carriage, T53. The proposed military characteristics were tentatively established as follows: *1. Purpose of this development was to provide a highly mobile 90mm gun incorporating maximum anti-tank fire power with mobility and protection for the crew and utilizing a maximum of components already in production. 2. Type – Full track based on medium tank chassis. 3. Crew of 7 men. 4. Physical characteristics: Weight not over 60,000 lbs. Length approximately 254". Width 104". Height approximately 110". Ground clearance 17".*

The armament would consist of one 90mm Gun M1 with 360 degree traverse with an elevation range of -5° to +30°. A cal. .50 ring mount was also to be provided, as well as provisions for an M1903 rifle with grenade launcher, three cal. .30 carbines M1 and one cal. .45 submachine gun. A radio set SCR-510 was also to be mounted. This was officially approved on 30 July 1942. Chrysler, however, was not waiting around. The pilot model was built before this, and shipped from the factory to Aberdeen on 02 August.

The 90mm Gun Motor Carriage, T53 consisted of a modified M4 medium tank with a standard 90mm AA gun, less out-rigger base, mounted on the rear of the chassis. No modifications of the gun or pedestal were made. The tank chassis was modified for this purpose by removing the turret and installing the Wright Whirlwind engine in what was formerly the front part of turret compartment. An ammunition rack holding 20 rounds was constructed between the tracks underneath the gun platform with heavy armored doors opening to the rear. Folding platforms were provided around the gun that could be lowered when the piece went into firing position. In travelling position these platforms were raised and formed a light shield around the gun pedestal. No modifications were made in the driving compartment. An SCR-610 was located in the right sponson.

An initial report within a week of receipt indicated T53 was a fine anti-tank weapon, but also indicated that whoever was being told that it was also a capable anti-aircraft weapon was, perhaps, not being entirely well informed. Following the demonstration of the pilot 90mm vehicle at Aberdeen, a conference was held at Headquarters, Services of Supply on 24th August 1942 attended by representatives of Services of Supply, Army Ground Forces, and Ordnance Department. At the conference it was agreed that 500 90mm Gun Motor Carriages, T53, with certain modifications should be procured during 1942. These modifications were:

a. Relocate the gun and carriage forward at a position substantially over the centre of gravity of the vehicle.

b. Relocate the engine to the rear of the vehicle at approximately the position occupied in the tank chassis

c. Provide a shield of bolted construction so that the entire shield, the front or the top might be removed in the field if desired.

In order to provide a GMC incorporating the above modifications, and also to study the possibilities of mounting the 90mm gun so that it might also be used as an anti-aircraft weapon, two additional pilots were constructed and designated as 90mm Gun Motor Carriage T53E1.

Contracts were placed with Allis-Chalmers (100), Watertown Arsenal (12) and Worthington Pump Company (280) in Sept 1942 for gun carriages suitable for mounting on the vehicles, with deliveries expected November and December of that year. If no remote control equipment was to be mounted on the vehicle, the gun could be depressed to -5 degrees. In the interest of haste the carriages were going to be of the standard

90mm mount M1A1 type, with the expectation of a new mount design after this. It was considered that a higher trunnion on models after #500 would allow a greater depression. Similarly, a new sighting system was expected to be used, but in the interim, the standard elbow telescopes would be used.

Tank Destroyer Branch wanted its guns to have armour protection for the crew. This series of photographs, dated 21st August 1942 shows the first proposed gunshield design. This was apparently deemed excessively ugly, as a second was also proposed, see next page.

After almost a month of testing, T53 was shipped to Temple, Tx, for evaluation by the Tank Destroyer Board. Results of this testing indicated that T53 was not preferred over the 3" Gun Motor Carriage M10, but that the 90mm was a good idea and that development of a 90mm gun motor carriage be continued. More specifically, the Tank Destroyer Board found as follows:

- The gun appears to be extremely accurate, and the gun platform is believed to be satisfactory.
- The high silhouette makes the vehicle difficult to conceal in open terrain. However, the broken lines of the gun and upper carriage lends itself well to concealment in woods or among scattered trees. The forthcoming installation of a gun shield is expected to make concealment more difficult, but a shield is considered necessary for combat.
- The lack of protection to the gun and crew makes maneuvering in woods or high brush hazardous.
- Low depression angle of the gun, minus 5, limits the employment of hull defilade.
- The vehicle lacks the speed desired in a Tank Destroyer.
- The vehicle lacks satisfactory stowage facilities.

- There is no adequate provision for the crew when riding on the vehicle. The necessity of lowering the folding platform and opening the doors of the ammunition compartment unduly increases the time required to put the gun into action, the best time recorded being approximately 35 seconds. However, by limiting the traverse of the gun to approximately 40°, the gun can be fired without lowering the side platforms. This shortens the time required to go into and out of action to approximately 10 seconds provided a supply of ammunition is carried on the gun platform. No racks for this purpose are presently provided. To require the ammunition passers to dismount and open the ammunition compartment entails a delay. The gun has been successfully maneuvered with the rear platform lowered.

- The location of the ammunition compartment is objectionable both from the point of service of the piece and because of the heat transmitted from the exhaust pipes which run alongside the ammunition compartment.

- The lowering of the rear platform makes access to the ammunition compartment difficult. Two ammunition passers are required – one operating from underneath the rear platform removes the round from the ammunition rack and passes it to the second passer who in turn passes it to the loader."

Above: The far less hideous, but still very prominent second gunshield proposal, also photographed 21st August 1942

Doubtless a test evolution for which there were many volunteers, T53 was subjected to four drag races against medium tanks. Over the 2.5 mile course, the conclusion was "There was no material difference in speed". All photos taken at the finish line, 22nd October 1942

On a more tactical level, the above photograph shows the relative silhouettes of T53 compared to an M3 Medium and M10 GMC. Of course, the gunshield must be added to this. Another reason to add the gunshield was demonstrated by the photograph on the right, taken to illustrate the hazards to crew and equipment (such as optics) posed when travelling through brush when unprotected by a shield.

Top: Detail of the gun mount
Above: The ammunition compartment doors, open and closed.

Both above: Demonstration of the crew positions on T53. Note the requirement of ammunition passers, and their uncomfortable-looking posture

By November 1942, the idea of T53 being primarily an anti-tank weapon with secondary AA capability had been discarded, and by direction of General Barnes, the second vehicle was to be an anti-aircraft project, with outriggers and a radio-control terminal at the rear. The Tank Destroyer Board recommended that the development of a dual-purpose 90mm GMC on a tank chassis be discontinued.

A further report from the Tank Destroyer Board also arrived in late November, furthering the end of the T53 as a Tank Destroyer, in which it declared the vehicle obsolescent given the development of T71. Further testing made clear further deficiencies, adding to the list from the earlier test:

- An 8-man crew was required to be effective.
- Because a man had to be under the rear platform in order to access the ammunition stowage, the vehicle could not operate on terrain that did not allow him to stand there, such as an upslope which reduced his space available.
- Some 40-50 rounds capacity were considered necessary.
- The muzzle-forward center of gravity of the gun meant that traversing with the vehicle at an angle could be a two-handed operation. This was a problem for properly tracking a target when the gunner also had a separate handwheel for elevation.
- The 90mm round demonstrated an increase of only a ½" of penetration as compared to high velocity 3" rounds then available.

Left: Stabilizing outriggers were added when it was decided to better support the AA role.
Right: Overhead view of T53, Oct 1942.

For tactical testing, T53 was given to a tank destroyer battalion in place of one of their M3 half-tracked gun motor carriages, with results more or less in line as could be expected: The M3 could use bridges that T53 at its 65,000lb weight could not, but the full-tracked vehicle could cross obstacles that the half-track had to detour around. On a more general basis, it was considered to be some 12 tons heavier than the Tank Destroyer branch wanted in its vehicles, and it wasn't any faster than the tanks. The

Left side view of the T53 with the gun in travel configuration.

Tank Destroyer branch decided in the end that it had no interest in a dual-purpose (and it underlined "dual purpose" in the report) 90mm gun carriage of any sort.

Two views ofT53E1, the left photograph taken 29 January 1943. Note the outriggers and travel lock. The swastika scratched into the turret side is unexplained. The detail photo on the right is dated 13 February 1943.

T53E1 arrived in Aberdeen January 12th, with a third vehicle, with a redesigned shield to reduce weight, not far behind and immediately sent on to the Anti-Aircraft Artillery Board at Camp Davis, North Carolina. The vehicle consisted of a Medium tank M4A1 chassis on which a 90mm anti-aircraft gun Ml had been mounted. Most of the principal components of the 90mm AA Gun Mount MlAl, including the remote control system and the leveling mechanism were used. Four folding outriggers with jacks at the outer end were attached to the chassis of the vehicle for firing stability. One-half-inch shields were attached to the traversing parts of the gun for protection of the remote control system and operators from small arms and flying debris.

A report from Aberdeen dated 6 August 1943 found that T53E1 was comparable as a firing platform to the standard 90mm gun mount on hard or frozen ground, but slightly inferior to it on soft or muddy ground. It was recommended that with certain modifications included in the report the vehicle be considered a satisfactory self-propelled mount for the 90mm AA Gun.

Top: T53E1, first pilot, as seen on New Year's Eve, 1942.
Above: The same vehicle as seen 13 February 1943 in firing condition.

Above: Two more views of T53E1, vehicle #1.

Below: The gunshield of the second T53E1 was more rounded and smaller.

However, the Anti-Aircraft Artillery Board disagreed after its tests, concluding that the gun did not meet the requirements of AAA and that the combination of the 90mm gun and the Medium Tank M4 did not show sufficient promise for use as an anti-aircraft weapon. They found the service of the piece difficult, and the way the shell casings were ejected everywhere (including at driver's head at some angles) combined with tripping hazards made the it "extremely hazardous to personnel." The AAAB also preferred a towed mount, as the immobilizing of a towing vehicle would not immobilize towed ordnance in such a case. This resulted in a recommendation in May 1944 that the projects T53 and T53E1 be terminated and that the vehicles be retained at Aberdeen Proving Ground for the historical record.

This recommendation was officially approved in OCM 23926 of May 24th 1944. It is worth noting the number of vehicles so disposed which no longer exist.

Top: Two photos of the second T53E1, factory fresh on 29th December 1942.
Above, left. The vehicle set up for firing. Above right: Photograph taken to demonstrate the hazard to the driver of ejected shell casings falling into the front hull. A casing is visible just right of the transmission.
Below left: Setting the fuze required navigating the engine deck. Note the heavily greased breech block.
Below right: T53E1 second vehicle as emplaced for firing.

There were more photos available of T53E1 than there was text to associate with it! Clockwise from top left: Two more photos of T53E1 second pilot in travelling configuration with travel lock engaged, outriggers and platforms folded. Photograph demonstrating the hazard of shell casings ejected onto the narrow space between the mount and the engine compartment. T53E1 was fitted with the ability to install sights for indirect fire. Another factory study of the second pilot in travel configuration. The vehicle in firing configuration.

T53 was never really supported by Bruce: He saw it as a distraction from the T70 he felt TD Branch needed.

90mm Gun Motor Carriage T71 (M36)

T71 Pilot #2 at Camp Hood

By September 1942, investigations by the Technical Division of Ordnance Branch indicated that it would be possible, with limited modification, to place the 90mm anti-aircraft gun and ammunition into tanks and gun motor carriages then using the 3" gun. Two candidate vehicles were particularly identified, Medium Tank T20 and Gun Motor Carriage M10. 90mm guns so modified would be designated 90mm Gun T7. OCM 19055 of October 22nd gave approval to the diversion of two guns for such testing.

After a demonstration firing at Aberdeen of a 90mm gun in an M10 in December 1942, the general idea was validated, although there were problems with the recoil system and the ejecting of shell cases. In January 1943, the Subcommittee on Automotive Equipment reported to the Ordnance Committee that the inclusion of the 90mm gun "appears to be a very simple problem of substituting the larger gun in the 3" Gun M7 cradle. […] It is therefore considered highly desirable to develop the possibilities of the 90mm Gun on the 3" Gun Motor Carriage, M10 chassis in order that the potentialities of this heavier weapon may be utilized and to have on hand a weapon (or the data necessary for rapid change to this weapon) that will cope with any vehicles our enemies now have or will have by the time this weapon can be produced". The mission statement was "To provide a highly mobile 90mm gun incorporating armor protection for the crew and utilizing a maximum of components already in production." OCM 19845 of 4 March 1943 gave the program the designation T71.

Tank Destroyer Branch was not impressed, and a LTC Shaffer filed a qualified concurrence in the program "with the understanding that this project is a development project only for the purpose of securing information with regard to the practicability of mounting the 90mm Gun on the Gun Motor Carriage, M10. The gun is not desired by the Tank Destroyers as a tank destroyer weapon since it is believed that the 3 inch gun has sufficient power. It is further felt that the Gun Motor Carriage M10, is too heavy and too slow."

The Chief of Engineers chimed in with his own concern:

The proposed vehicle exceeds the limitations as now prescribed by AR 850-15 in weight, width and gross weight per lineal foot of ground contact, It is, however, within the limitations of the proposed revision to paragraph 5, AR 850-15 for a special purpose combat vehicle assigned an armored division or designed to accompany an army in the field.

The Chief of Engineers concurs in the manufacture of two of the proposed vehicles if they are not to be assigned to an infantry or cavalry division, but desires to bring to the attention of the Secretary of War the following information:

Using a ground contact length of 12' 3", as contained in the physical characteristics for the subject vehicle, the maximum permissible weight under AR 850-15 is 31,350 pounds. The maximum permissible weight under the provisions of the tentative proposed revision to AR 850-15 is 67,925 pounds. The proposed gross weight of the vehicle is 62,008 pounds. Therefore, the vehicle exceeds the limit specified in AR 850-15 at present but is within the limitations of the proposed revision.

The American Railway Engineering Association loading diagram indicates that this vehicle may be transported over many railroads conforming to these requirements which include almost all United States railroads. It cannot be transported on many railroads in England and some railroads in other foreign countries. This fact should be checked by the Transportation Corps.

A telephone conversation on 13 March 1943 between COL Robins of Army Ground Forces and MG Barnes further indicated the lack of emphasis on the program, with the statement that Ground Forces had no interest at all in the 90mm M10. MG Barnes responded that it was better to try these things before they were needed, and it was better to have the experimental work already done. At this point, the 90mm M10 was about to become a victim in the Army's tendency of the time to place, and then cancel, large orders for vehicles, with the result of expensive cancellation charges. AGF had no interest in adding another vehicle to the list of budget disasters, and there was a directive from General Syter that any projects not essential to the winning the war were to be reduced. MG Barnes apparently took exception to what he believed to be an intrusion into his territory, explaining that if anyone was going to cancel one of his projects as unnecessary, it would be him. MG Barnes then hung up on COL Robbins, called General Woods, and discussed the matter with him. Woods pointed out that General Somervell and General Lucius Clay, Director of Material, were both quite concerned about the incessant changes of mind on the development and purchase of equipment, but MG Barnes did succeed in maintaining T71's development program.

The mock-up of the T71, which placed a wooden turret on an M10's hull

Although the T7 gun had been successfully tested in the M10 by this stage, a true T71 had not yet been built. Chevrolet indicated that if the Ordnance Dept gave them (at no cost) 2x M10 carriages, 2x 90mm T7 guns, 2x sights and 2x turret traversing mechanisms, they'd build the two pilots. The contract for this was signed after about two weeks' deliberation, with the first wooden mock-up being completed within a week, on 10 May. However, the contract was later transferred to Ford Motor Company in July of that year.

Come September 1943, with development well along but before the first test report of T71 at Aberdeen Proving Grounds, a document was submitted to Army Service Forces from the Acting Chief of Ordnance (MG Hayes) outlying proposed requirements for higher-firepower vehicles. In addition to the 1,000 76mm M4A3 proposed, and the suggested ordering of 500 both of tanks T25 and T26, it declared that 500 Motor Carriages M10 with 90mm High Velocity gun should be authorized immediately, with deliveries expected in early 1944. This wasn't a new concept, proposals to replace M10 in production with 500 T71 had been initially put forward by Tactical Development Branch the month before. In fact, such was the changed push now from Army Ground Forces that a month later, they submitted another document suggesting that production of M10 and M10A1 be ceased in favour of T71 immediately, no matter what the cancellation charges would be. They also expressed a preference for the Ford engine. The response to this was that the British orders for M10 were sufficient to cover all M10 contracts currently in place, so no worries cancelling the American requirements.

Repeated requests to authorize 500 vehicles were made through September and October, and denied, presumably as part of the reluctance to commit to unproven vehicles. General McNair specifically stated on 1st October that he did not want any self-propelled mounts for the 90mm at all, and this included the tanks, as they were too heavy. His opinion did not hold sway. Repeated pressure from Army Ground Forces and Technical Division resulted in an acquiescence by Army Service Forces for an order of 500. By November 4th, MG Clay was looking at the 533 M10A1 scheduled for production by the end of the year, and suggested simply that the vehicles be built exclusive of the turrets.

The test report for T71 Pilot #1, which shipped the week of 6 Sept 1943, came back from Aberdeen Proving Grounds 20 October 1943.

T71 Pilot #1 at Aberdeen, left-rear quarter. All Aberdeen photos taken 25 Sep 43

This vehicle was basically a 3" Gun Motor Carriage M10A1 carrying a 90mm gun cradled in a 3" recoil system; breech, semi-automatic; trunnions positioned so that gun unloaded had a slight muzzle preponderance. This was mounted in a specially designed turret which weighed 11,970lbs unstowed. The turret had a cast bustle on the rear for the stowage of 12 rounds of ammunition; a Cal. .50 machine gun ring in the left rear corner in place of the old Cal. .50 fixed mount. A semi-basket suspended from turret ring carrying a seat for the gunner right of gun, a standing platform and folding seat for the gun commander was directly behind gunner. The loader was not carried by the semi-basket but stood on the floor of the crew compartment. And an Oilgear power traverse was installed. Other changes were the addition of a 750 watt single-cylinder gasoline powered auxiliary generator in the engine compartment and the addition of a slip-ring box in the center of the fighting compartment floor.

T71 Pilot #1 at Abereen. Note the large ring mount for the Cal. .50 (And no, the photograph is no longer restricted!)

The turret received particular praise for being so well balanced that less than 3 lbs of effort on the manual handwheel was used to traverse it on level ground, and 10 pounds on a 30% slope. By way of comparison 90mm GMC T53 could require 140lbs of effort to traverse. As with any new vehicle, there were some tweaks necessary. One was the fact that feet kept getting caught between the gunner's platform and the hull, with two persons suffering injury as a result: It was recommended that a 6-8" lip be placed around the edges. The only other recommendations were that the ready round ammunition racks be modified to make extraction easier, that the azimuth indicator be ventilated to allow moisture inside it to evaporate, and placing a guard to protect the intercom wiring. Provided those recommendations were made, as far as Aberdeen was concerned, T71 was to be recommended as satisfactory.

The next organization to test the vehicle was the Armored Board, who subjected it particularly to a battery of firing tests in December and January. Their initial report was submitted 14 Dec 1943, and made a number of suggestions:

-The ammunition stowage racks in the sponsons are not large enough to stow ammunition without removing the fiber cap. The racks are 39 inches long and the ammunition within the fiber case is 39-1/16 inches. The racks must therefore be nearly 39-½ inches long.

-The anti-aircraft ring mount interferes with the service of the piece. It also is so poorly positioned that firing Cal. .50 anti-aircraft fire from this mount is difficult. The ring mount should be replaced by a pedestal mount in position to permit firing the Cal. .50 machine gun from within the turret, permitting 85 degree elevation for a limited angle, and also where it can be manned from the deck of vehicle. The new equilibrated cradle should be used.

-The position of the hand traverse unit cramps the gunner. Moving this unit approximately 1-½ inches to the right would relieve the gunner's cramped position

-Stowage arrangements should be provided for a Barr and Stroud Range Finder, M9.

-A fifty yard extension cord should be provided for the tank commander's interphone connection to permit observing and directing fire from the flank.

-The impulse relay for the firing solenoid should be modified by changing the position of the cable connection from the present position in rear to one at the bottom of the relay to facilitate disassembly of the breech. As it stood, the gun needed to be jacked out of battery in order to disassemble the breech.

Above: T71's gunner's station.
Below: The reconfigured Cal. .50 pedestal and stowage, as seen on Pilot #2.
Bottom: Top view, Pilot #2

By the time of the report of 17 January, a second vehicle had been provided. This report, and the accompanying pictures, demonstrate the new MG mount which was on top of the turret bulge with a standing platform for the operator suspended from the turret ring with stowage brackets for spare barrel and ground tripod in the same location. The bustle stowage was by now reduced to eleven rounds, with only five now being on the loader's side.

They found that the 90mm was exceptionally accurate, and exceeded the capability of the sights. Reference was made to lead indicators in the reticle up to 4,200 yards demonstrating an expectation at that distance, assuming that they could see past the muzzle blast. A flank observer was required in all but the best possible circumstances in order to sense rounds. Accuracy in the firing tests were as follows for 5-round groups.

	Elevation	Deflection
500 yards	3"	3"
1000 yards	11"	3"
1,500 yards	12"	12"
2,000 yards	24"	18"
3,000 yards	63"	12"
4,000 yards	84"	33"

Obscuration of the gunner by muzzle blast was considered of particular concern, being significantl worse than the 76mm, though only slightly worse than the 3". Testing the 3" with a muzzle brake and long-primer ammunition indicated that this problem would be mitigated on the 90mm by similar methods. As a result, Armored Board considered that the muzzle brake and long-primer ammunition be given "the highest priority." Further, both because of signature issues and the obscuration problem, the report noted that "The extreme blast obtained from this gun makes it imperative for the first round to be accurate. This necessitates the using of a range finder to accurately determine the range. Therefore, provision should be made for the stowage of a Barr & Stroud range finder."

Padding was found to be necessary around the top edge of the turret to prevent injury to the crew. The position of the plumbing for the oil traverse unit was unsatisfactory as it was in a position to be damaged by the gunner's feet. The electric cabling from the finger firing switch to the firing solenoid interfered with the gunner. A headrest for use with the direct sight telescope should be provided which was shaped to the head to properly position the gunner's head, and accommodate both left and right eye dominant gunners, as the one installed was poorly positioned and suitable for right-eye use only.

Outside of that, and suggestions for a cover to stop dirt getting into the .50 cal pedestal, a bag to catch shell casings, and some reconfiguration of the .50 cal. stowage, the Armored Board had little to add.

T71 Pilot #2 as received by Tank Destroyer Board, Camp Hood, TX.

The Tank Destroyer Board at Camp Hood was the next to chime in on February 1944 with the results of its test of pilot #2. Amusingly, it was not happy with the pedestal mount for the .50 cal., after engaging both balloons and radio controlled aircraft with it, and requested a ring mount be fitted on the gun commander's position. A 500-mile test indicated that the flat rubber tracks that the vehicle came with needed to be replaced by steel or chevron rubber, as it had limited traction on mud. Outside of that, no difference was noted between the operation of T71 and M10A1.

Above: Loader's station
Left: T71 Pilot #2 after a little careful use, on 9 Feb 1944.

TD Branch's gunners, however, did agree with their Fort Knox counterparts that the position was cramped, and the loaders that the secondary stowage in the hull was inconvenient to access. They also engaged targets at up to 4,000 yards, with the following observations:

Firing tests/were conducted on moving targets, 10 to 20 mph, at 500 to 1,700 yards. Results showed that effective fire could be placed on moving targets using power or hand traverse. However, rapidity of fire was adversely affected by target obscuration due to smoke. On the plus side, the dust element was negligible due to continued rains. The gun also showed a marked drift to the right which, at 3000 yards, required a 2-mil correction and at 4,000 yards, 2.5 mils. It was recommended that a new reticle be created which included this curved deflection line which accounted for the drift as determined by the Ordnance Board. It was also recommended that a x5 telescope was necessary for engagement beyond about 3,600 yards, whilst noting that no acceptable one actually existed, T92 being unsatisfactory.

Capability with HE rounds was also tested, with foxhole targets engaged by HE rounds with delays from 0 to .15 seconds. Ten times as many simulated casualties were struck by the delay rounds, as the rounds tended to impact up 5-15 yards before detonating, causing an airburst. The only 'casualty' caused by quick fuze was hit by the shell itself. Muzzle blast at night was found to be far larger and brighter with HE than with AP.

Left: The impulse imparted onto the hull from firing can be seen by the tracer panel
Right: This photograph of the mired T71 Pilot #2 indicates the level of careful use
given to the vehicle by Tank Destroyer Board during testing

In the end, the Tank Destroyer Board report concluded:

The 90mm Gun Motor Carriage T71 is suitable as a primary tank destroyer weapon.
The T71 has capabilities due to its heavier caliber, which make it superior, especially for secondary missions involving shock action against concrete, to the M10 or M10A1.

Come May 1944 a new gun mount shield (E-10560) was provided to replace the 950lb, 1¾" shield mounted originally in the pilots. The Tank Destroyer Board mounted and tested this, concluding:

This heavier shield introduces problems in this gun motor carriage. The shield itself adds 450 lbs., and if a balanced turret and gun mounting is to be had, approximately 300 lbs. counterbalancing weight is required, 200 lbs. on the turret bulge and 100 lbs. on recoil guard. Any deadweight is objectionable in the gun motor carriage, but the heavier shield appears tactically necessary in order to obtain 3" armor basis at the front of the turret. This heavier basis appears justified by the probable employment of this gun motor carriage for an attack of fortified positions using direct fire from hull-down positions.

Pursuant to these and other recommendations, further tweaks to the design were made, with a new equilibrator being designed to balance the guns when fitted with a muzzle brake and new sponson stowage being created. This new configuration was tested in July at the General Motors Proving Ground. Still, a heavier elevating mechanism was required, and larger springs for the new equilibrator were also needed. However, this was not going to stop delivery of M36 overseas without the muzzle brake. (It should be noted that though the device actually is a muzzle brake, the desired effect seems to be of "blast deflector," with no mention made of recoil problems.)

The prototype of the revised gunshield.

Above: Rear view of T71 Pilot #2 as received by Camp Hood. Below: A canvas "roof" was provided for protection against the elements. This shows the supporting bows installed.

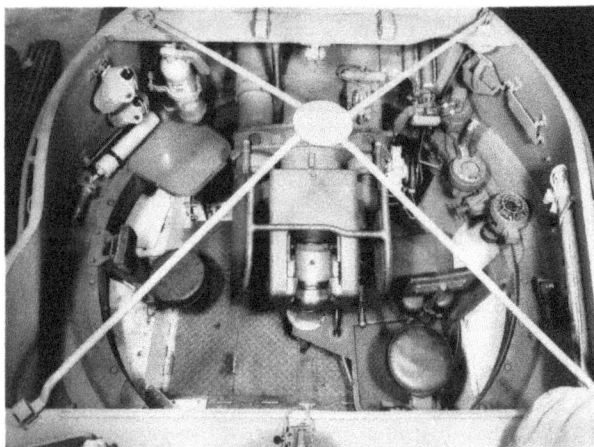

On 1st June 1944, the Ordnance Committee recommended standardization of T71 as Gun Motor Carriage M36, a recommendation approved 22 June 1944. As an aside, by this point Ordnance Dept. was on a bit of a naming kick, and had recommended the month before that T71 be given the nickname "Black Jack." The first 300 90mm guns were being manufactured by Fisher, with Massey-Harris manufacturing the remainder.

Shortly after the invasion of Normandy, the need for the vehicle was re-evaluated and given something of a more pressing priority. The urgent requirement for GMC M36 and Medium Tank T26 (Heavy Tank T26E1) was set out by the Director of the Production Division, Army Service Forces in a memorandum dated 10 July 1944, and a meeting in Detroit the following day laid out plans for increasing production of M36 and also converting 187 M4A3 medium tanks into gun motor carriages in order to make up for a shortfall in production capability. This conversion consisted of removal of the Medium Tank Turret, Gun Mount and basket, revision of interior storage and ammunition arrangements and re-installation of the turret group from the 90mm Gun Motor Carriage M36. Minor exterior and interior hull changes were also required. A SCR-610 would be included, a redesigned gun travel lock provided, but the bow-mounted cal. .30 machine gun retained. It was thus recommended to the Ordnance Committee on 18 Sept. 1944 that Carriage, Motor, 90mm Gun M36B1 be classified as a Required Type, Adopted Type, Substitute Standard Article.

Although development work had been deemed complete in October, further work did continue. Armored tops were developed for all the tank destroyers in January 1945, with them being added to the standard production models by April. The other significant development was the introduction into service of M36B2, made by converting 3" Gun Motor Carriages M10. This was officially read into to the record on 22 March 1945, though the actual conversions did not start happening until May.

Left: Photographs of M36B1 under test seem hard to find. This photograph is taken from the operator's technical manual. Right: The eleven rounds of 90mm stowage in the bustle, behind protective canvas bags.

Four photographs from Camp Hood showing the roof fitted to M36s. The roof had a number of liftable panels for the panoramic sight and visibility, as well as required access.

Finally, the nickname. By the time of Gen Barnes's order of 24 November, possibly because it may have effectively given Gen Pershing two vehicles, the name "General Jackson" had been selected over the previously proposed "Black Jack".

M18 with 90mm GMC M36 Turret

The experimental 90mm M18. Not being an official Ordnance project, it was never given a designation.

In late 1944, a letter from Twelfth Army Group put to Army Ground Forces a series of improvements they desired of the M18. These included greater frontal armour, a roof, a coaxial cal. .50 machinegun, and a more powerful cannon. Tank Destroyer Board was directed by AGF to initiate a study of these proposals. The result came back in January 45:

The ideal requirements [...] means a vehicle of basically different design from the M18. If all are met, these requirements will result in a vehicle of entirely different employment characteristics, - in effect a tank. Such a vehicle would fail to meet the desires of Third Army as to mobility.

Still, they dug into it and concluded that the addition of thicker armour would give no practical benefit at the cost of slowing down the vehicle and adding wear, a suitable roof had been developed, a coaxial machinegun was possible, and that a more powerful gun couldn't be mounted for two reasons: Firstly, that Ordnance had declared the 76mm to be at the limit of its pressure capacity, so no improvement could come from that quarter, and secondly, that installing a 90mm gun into the M18's turret would require significant redesign. Fortunately, the report also concluded that given that in practical experience the M18s weren't coming across much that they couldn't kill, but which the 90mm could, it wasn't considered that important to upgun the vehicle. There was, though, a caveat: "However, in anticipation of exigencies as the war continues and of possible resumption of the manufacture of the M18, which is a proven vehicle with no counterpart as a combination of fire-power and mobility, it is believed advisable that steps be taken now to explore from the design viewpoint the possibilities of replacing the 76mm gun on the M18 with the 90mm caliber, especially any improved versions of that caliber."

The report did put forth a further consideration: "To convert the present considerable stock of unassigned M18s to 90mm vehicles would require, after development, experiment and test has been completed, an unpredictable number of months. This would tie up this stock of an admittedly efficient weapon. [...] The advisability of tying up completed, capable vehicles actually ready for combat assignment appears questionable." The upshot of it all was that the idea of mating a 90mm turret from M36 to M18 was initially investigated in early February 1945 at Aberdeen, and measurements taken. A note from COL Montgomery, President of the Tank Destroyer Board on 14 March indicated the desirability of actually trying it out, but that there was no pressing need. The response, dated 15th March said "We will try this conversion as soon as the old M36 is no longer needed for traverse and other directive tests"

That took perhaps a little longer than anticipated. It was 25 July before Gunnery Section was finished with its M36 and Ordnance could set about taking the turret. As a matter of some interest, the memorandum records on the matter clearly express that the modification occurred in July, but the photographs of the installed turret are dated 27 June. It is the author's opinion that the photographs are incorrectly dated.

The one and only M18 with the 90mm turret

The actual installation of the turret "was accomplished with relative ease", the modifications required were:

-The half floor of the turret was raised approximately two inches by cutting two inches out of the support members which fastened the floor to the turret ring. This necessitated relocation of the conduits from the collector rings to the traversing mechanism and to the radio.

-The slip rings (from M36) and yoke (M18) were raised approximately two inches by simple modification of the yoke bracket which consisted principally of drilling two 5/18" diameter holes and a few minutes' work with a hack saw.

-The discharge air vent from the converter cooler blower interfered with traverse of the turret. It was necessary to cut this vent back and reduce the discharge cross-sectional area.

With the turret in certain positions, the drivers' hatches would not open, and it was necessary to trim the smaller hatch door two inches in order for them to open in any turret position.

Once the 90mm turret had been installed, without stowage, the vehicle weighed 43,075lbs, approximately 3,000lbs more than the standard M18. Although the overall performance of the vehicle was not noticeably affected, the shift in the center of gravity resulted in the vehicle assuming something of a nose-high attitude.

Test firing of the cannon both with and without the muzzle brake indicated very definitely that the muzzle brake should be used. Without one, the vehicle rolled 22" when fired to the front without brakes applied, and rocked excessively to the side. With, the rollback was all of ¾". Shot groupings at 1,000 yards was 12"x7" to the front, and 10"x10" to the side.

The vehicle was then immediately set about to driving on a 1,000 mile endurance run. In order to mitigate the increased ground pressure caused by the heavier turret, the M18 was fitted with 21" T82 tracks. These had been independently tested on an M18 and M24, with mixed conclusions on the question of if the increased floatation was worth the increased weight and rolling resistance. The answer was not to be found in this test: 25 miles into the test, the worn tracks failed. The test was then concluded with regular tracks, with no disastrous breakdowns.

This and previous page: The photographs clearly show that the M18 that was used was one of the few to receive a two-tone camouflage scheme

a. The installation of the turret can be accomplished in the field, if necessary
b. The turret functioned satisfactorily on this chassis when the gun is equipped with muzzle brake
c. The vehicle stood up satisfactorily in the 1,000 mile endurance run.

The recommendation was "that this installation be considered satisfactory and worthy of further development"

Of course, by the time all this was done, the war in Europe was over, and the war in the Pacific was within its last couple of weeks. The final report on the project was dated January 1946. As a result, there was no particular desire to re-turret the M18s, and the experiment stopped there.

90mm Self-Propelled Anti-Tank Gun M56

The idea of developing a self-propelled high-velocity, small caliber gun which could be transported by air and used to hold airheads against enemy armour was first presented at an anti-tank conference held at Fort Monroe in October 1948. Following investigation of the problem, in December 1948 the Army Airborne Panel established a requirement for such a combat vehicle. The required military characteristics for the new item were formulated by Army Field Forces and on 27 April 1950 a project for the development of the T101 self-propelled gun was opened by Ordnance. After several months of negotiation, a contract was awarded to the Cadillac Division, General Motors Corporation in October 1950 to design and build two pilot vehicles.

Although the project got off to a slow start in 1950 because of lack of funds and the higher priority given to the tank program, development subsequently proceeded without any unusual problems, despite several major changes in the original military characteristics. At first, for example, it was planned to mount the T119 (M36) high velocity tank gun as the armament of the new vehicle. But it soon became apparent that too many design changes would be required before the gun could be satisfactorily mounted. Accordingly, a project was opened for the development of a new high velocity 90mm gun, designated the T125, specifically for use in the T101 vehicle. The new gun was designed to have the same ballistic characteristics as the T119. As another example, the original plan to design the new gun for transport by air in aircraft of 16,000 pounds capacity was abandoned in favour of making it droppable by parachute. The changes in design necessitated by this decision were minor, the chief of them being a reduction of over-all height.

All photos this page and next: The first pilot of T101. Note the large gunshield with two doors, and angled wings. These would not last.

Tests of the armament in 1952 indicated the basic fragility of the system. Four rounds were fired (in a 4' grouping at 1,000 yards) before the first breakage occurred. The front axle was measured as raising 19" upwards in recoil when the gun was fired level, and the elevating screw jammed. Aberdeen were able to repair it the first time, but after the second, a replacement had to be obtained from General Motors. Even at that, weights had to be added to the gunshield in order to make the system more muzzle-heavy, to reduce the upwards recoil tendency. Then the blast deflector failed. By the time over a hundred rounds were test fired, and a modified T70 mount added to Pilot #2, a report dated October 1953 described the weapon as "surprisingly accurate." In 1953, the weapon was shooting a 15-round group in a 40"x55" area at 1,000 yards using HVAP. It is worth pointing out that a 40lb aluminium shield was only to protect the crew from the blast of the gun. The steel gunshield weighed 365lbs.

The improved gun mount, T70E1 saw the primary cylinders relocated internally concentric with the secondary cylinders, but was otherwise pretty similar, this was done primarily to aid in the air-dropping configuration. Preliminary testing involved firing 589 rounds, so apparently things worked better. However, at least nine designs of muzzle brake and blast deflector were tried out, varying from four-port to T and Y shapes, none were successful. A 1955 report suggested that the counterweight be used until a better muzzle brake was designed, but it would seem that such a muzzle brake never materialized. Getting the hop down to acceptable levels was never to be obtained, with guidance being provided that in action, only the gunner and driver remain on the vehicle; the loader and commander should dismount. One comment in the test reports indicated the desireability of additional padding around the gunner's telescope eyepiece, but in practice, the gunner would frequently pull his head out of the path of the sight when firing. This led to a significant problem in adjusting onto a target, as the gunner would be incapable of sensing his own rounds.

T101 Pilot #1 accumulated over 14,000 miles in testing, and a number of deficiencies with the suspension were noted. The tracks consisted of 44" sections of dual-rubber strip connected by steel grousers. The bolts holding these bands together would occasionally shear off. Worse, the vehicle was very prone to throwing track. A number of improvements were made, most notably, squaring off the profile of the tyre, and reducing the outwards lean of the guide horns to ten degrees instead of the original twenty, these were later applied to Pilot #2 and tested in October 1955. Although an improvement, a new track still had to be designed.

In the meantime, further tests of T101 were being conducted in Arizona for desert operation, Colorado for altitude, and drop tests from a C-119 in Fort Bragg. Desert mobility was 'excellent', but 20-40% side-slopes resulted in thrown tracks. The big problem was that of dirt and dust thrown up with the vertices drawing the dirt forward onto the crew and even in front of the driver to such an extent that he had to stop. A number of solutions were trialled, to include redesigning the gunshield, extending the sandshields, and moving the exhaust. The lack of vision to the driver's right was rectified by telling the driver to stand up and look over the gun when wanting to look right. As the report said, "there does not appear to be any practicable redesign of the vehicle which could correct this condition." In order to get the vehicle into the C-119, 18"x12" sections had to be cut away from the top corners of the gunshield.

The sum of approximately $2.5m was spent on the development and initial testing of M56. Arbitrarily using the date of October 1955 when the vehicle was transferred from the R&D Division to the Industrial Division for product engineering as marking the end of the development phase, it will be noted that a period of 83 months elapsed from the inception of the project until the time the transfer was made.

The timeline was more or less as follows:
6-7 October 1948: Requirement established at Fort Monroe.
13 April 1949: Military characteristics prepared and forwarded to G4 on 12MAY1949.
12 May 1949: Army Field Forces recommended to G4 that the project be initiated.
6 June 1949: G4 requests Office of Chief of Ordnance to start.
20 October 1949: OCTM 33079 approved for development of gun and ammunition
27 April 1950: OCTM 33240 approved for construction of two pilot models.
June 1951: Contract signed with Cadillac Division, GMC.
May 1952: Pilots 1 and 2 arrive at Aberdeen Proving Grounds for Tests.
June 54-June 55: User tests completed.
October 1955 Transferred from R&D Division to Industrial Division.

Although Ordnance estimated in late 1955 that the T101 would be ready for issue to the field in mid-1957, the first production models were not delivered to the troops until June 1958. The reason for this delay was CONARC's waning interest in a conventional anti-tank weapon (considered suitable for use only by airborne infantry troops) in light of the noteworthy progress being made in the development of the DART anti-tank guided missile system.

The issue was joined in June 1955 when the final test report on the T101 was received at CONARC headquarters. Following a review of the report, CONARC agreed that the vehicle met the stipulated requirements. However, arguing that DART, which was expected to be available by 1959, would be more effective, less expensive to produce, and have a wider application than the M56, CONARC recommended against its procurement. The Chief of Ordnance and the Chief of R&D on the other hand, contended that the M56 satisfied the urgent requirement for a highly mobile, lightweight anti-tank weapon for use by airborne infantry troops, pending availability of DART. Failing to receive support for their view, they requested that both weapons systems be re-evaluated before any final decision to shelve the M56 was made.

T101 Pilot #2, seen here in 1952, had a smaller gunshield, with the gunner's door removed

There the matter rested until November 1955 when the study, which indicated that the M56 and DART were complementary, rather than competitive weapon systems, was completed. Accordingly plans were set in motion for the procurement of 160 vehicles on the basis of 30 vehicles per airborne infantry (ROTAD) division.

These two photographs are from the Technical Manual, and show a stowed production M56

Cadillac's satisfactory performance in bringing the M56 to the production stage marked it out as the obvious choice for the initial production contract. In November 1956 a preliminary purchase order was sent to the company, and on 21 February 1957 a contract for the procurement of 160 vehicles was concluded. The contractor undertook to deliver three pre-production pilot models by April 1957 with production models following from January to June 1958. Subsequently, this schedule was extended to August 1958. Unit cost was to be $107,000.

Upon receipt of a program directive from Office, Chief of Ordnance, calling for the procurement of an additional 291 vehicles, a supplementary contract was negotiated in April 1958, with delivery scheduled through May 1959. The unit cost for these additional vehicles was $43,333.

On its own initiative, Cadillac prepared a plan for a family of four more vehicles on the M56 chassis: An Armored Personnel Carrier, an ATGM carrier, a forward area anti-aircraft weapon, and a 105mm howitzer. Ordnance was not interested, and so Cadillac tried working directly with Armored Board and the Infantry School. Additional variants proposed now included a 106mm BAT (Battalion Anti-Tank recoilless rifle), one of which was built and demonstrated, an 81mm or 4.2" mortar carrier, and an amphibious cargo carrier. Needless to say, these didn't get anywhere, as the T113 program seemed to be proceeding well and would have been quite capable of equaling if not bettering the M56 chassis at many of these roles.

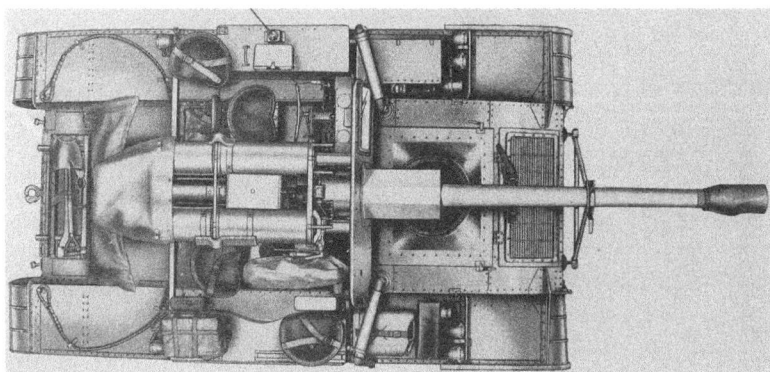

Two more photographs from the manual. The author was unable to locate any usable photographs of production M56 under testing.

105mm Gun Motor Carriage T95

Only one type of direct-fire 105mm gun motor carriage was made, and it was more commonly considered to be a super-heavy tank. However, for a while it was given a GMC designation (OCM 27219, reversed OCM 30958), and though not specifically designed to be a tank killer it was considered that it would be pretty good at killing tanks in the event that it should happen to find one in the sight reticle. So it is mentioned below for partially completeness' sake, but mainly because it appears to be a very popular vehicle and in case the author doesn't do a heavy tank book in the future.

Noting that the trend in tank design had been steadily toward heavier armour and more powerful armament, the Ordnance Department in September 1943 initiated studies of a very heavily armoured vehicle, with an 8-inch armour basis, mounting the new 105 mm T5E1 high velocity gun, and utilizing the electric drive system of T23. It was believed that this self-propelled gun would be of great value in attacking German West Wall fortifications because of the extremely good armour and concrete penetrating characteristics of the gun. Chief of Ordnance proposed that 25 such tanks be built, so that a sizable number would be available for use of the European forces should heavy fortifications be encountered, estimating that this number could be completed in from eight to twelve months.

At a conference with the Ordnance Department, Army Ground Forces recommended that only three pilot models be built, and a mechanical drive be used instead of the electric drive. In March 1944, Army Service Forces authorized the procurement of five Heavy Tanks T28.

The thickness and slope of armour was designed to make the vehicle very formidable and almost impossible to destroy in combat. Consequently, the gun was mounted in the forward part of the hull to provide a maximum of ballistic protection for all armour surfaces and the entire front of the vehicle, including the greater part of the track and suspension system, was protected by heavy armour placed at advantageous angles. Frontal armour thickness was increased to 12", though castings on both an 8" and 12" basis were manufactured by the General Steel Castings Corporation of Eddystone, PA, and sent for ballistic testing in late 1945. In this they were subjected to attack from 8.8cm and 90mm projectiles, also 16 and 40 pound demolition charges and 155mm explosive projectiles, generally performing well except in the area of the driver's hood bulge.

This page: Pilot #1, 23 January 1946

The vehicle was described in the 1947 test reports as follows:

The Super-Heavy Tank T28 is a heavily armored full track-laying vehicle with a 105 mm T5E1 gun mounted in the forward end. The gun can be traversed approximately 11° in either direction or a total of 22° altogether [Author's note: This is in contradiction to the test data listed further down the report]. It can also be elevated approximately 18½° and depressed 4°, allowing a total travel of 22½° in a vertical plane. The gun is elevated and traversed by two separate manually operated mechanisms. The vehicle powered with an 8-cylinder, V-type liquid cooled, 450 hp, gasoline engine, which drives the vehicle through a torque converter type transmission, a differential and individual final drives for each set of tracks.

Top left: The auxilliary brakes are identified in this photo of 1 May 1946.
Top right and above, two more views of T95 Pilot #1 on 23 Jan 1946

With the exception of the driver's vision cupola, the commander's cupola and the ring mount for the .50 Cal. machine gun, the vehicle has a very low silhouette with a relatively flat top. The overall width between the tracks, and the straight sides and low height, tend to increase the low appearance of the vehicle. This low appearance is enhanced even further by the rearward slope of the engine compartment. The muzzle end of the 105 mm Gun, when locked in traveling position, is the highest projection on the vehicle.

Sixteen sets of road wheels, individually sprung, support the vehicle on each side. The wheels are mounted in two sets of tracks on either side of the vehicle, each track set being protected by a heavy armor sloped plate over the top and a steel fender of 1" thickness which extends down over the outside edge of track to within 20" of the ground. With the aid of two manually operated hydraulic winches mounted on the vehicle, the outer tracks are removable as an assembly with the outer final drives intact, this being the most distinguishable feature of the vehicle. Both tracks on each side of the vehicle are driven from one final drive for each side, with a splined connection between the inner and outer track assemblies.

After removal, the outer tracks may be coupled together and towed as a unit. This greatly reduces the overall width of the vehicle for shipping purposes and facilitates its travel over narrow roads and bridges.

The hull is an all-welded structure of heavy armor, steel plates and castings. The fighting and engine compartment are separated by a water-tight pressure sealed bulkhead, which also acts as a fire wall and gives additional support to the roof plate. The power unit, consisting of the engine, transmission and differential is removable as one unit from the engine compartment.

This is the standard "power pack" used in Medium Tank T26, with the exception of having a greater final drive gear reduction. Two sets of final drives were supplied with the vehicle having reduction ratios of 12.126:1 and 10.62:1. The electrical system, a 24-volt single wire ground return system is constructed of waterproof conductors and terminals with specially constructed spring and rubber shock mountings for instrument panels and control boxes. Each circuit is assembled as a unit or individual harness assembly, lessening the danger of shorting or wire fractures. The cooling unit, which consists of four blower fans, two radiators for the engine cooling system, one oil cooler for the differential and one for the transmission, is removable from the vehicle as an assembly with the transverse housing. The fighting compartment is provided with a blower for slightly pressurized ventilation and the vehicle is equipped with only one set of driving controls. Auxiliary hydraulic brakes were also provided to aid in steering. Combat weight with crew of four men and equipment is 194,000 pounds.

This page: Photos taken in March of 1946 showing the creation of the 'trailer.' Note how the vehicle is jacked to match height

The testing process for T28 was somewhat hampered by the sheer size and weight of the vehicle. Ordinarily centre of gravity was determined simply by lifting the vehicle with a crane, but the largest crane at Aberdeen was incapable of lifting T28. As a result, they took the outer tracks off, and lifted the main vehicle and the tracks separately. Similarly, weight distribution measurements could not be taken as the scales weren't designed with such a wide vehicle in mind. Again, the outer tracks were removed, and calculations made (The average load on each pair of wheels being concluded to be 11,812lbs with both sets of track installed). The report estimated that the use of self-aligning bolts instead of the squared ones used in the vehicle could shave almost an hour off the 2+ hour time to conduct the installation/de-installation operation.

Testing was marred, however, when the muzzle brake was blown off by the second round fired and flew 500 feet forward. It was concluded that it was a metallurgical failure, not caused by the projectile.

Acceleration and braking proved to be stellar, taking only 8.8 seconds to achieve the vehicle's maximum speed from rest, and braking in 9.4 feet. Granted, maximum recorded speed was only 8.8mph. The paragraph entitled "Endurance operation" was prefaced by the note that "Accumulation of mileage on Super-Heavy Tank T28 took considerable time because of the slow

26 March 1946, an occupational hazard of testing new equipment

operating speed (5-6mph) and low priority assigned for maintenance requirements." Still, they clocked up 541 miles in the testing, with a fuel consumption rate of 0.16 miles to the gallon. Oil was consumed at a relatively more reasonable rate of 4.13 miles per quart.

Above left: An undated photograph of the front of the outer tracks in 'trailer' configuration, but unlike the earlier photograph, this time a tow-bar is used. This would certainly help reversing. Top right: Top view, 23 Jan 1946

Overall conclusions were:

a. The increase in armor and weight, without corresponding increase in power capacity, has critically reduced the mobility of this vehicle. From the standpoint of mobility, reliability and performance, the Super Heavy Tank T28 is unsatisfactory.

b. The 105mm Gun T5E1 is satisfactory.

Side view, of the 'inside' of the outer track system.

While all this was going on, on 15th April 1946 the subcommittee on automotive equipment sent the Ordnance Committee a letter. "[T]he Heavy Tank T28 was renamed as a 105mm Gun Motor Carriage T95 on the grounds that the vehicle, since the principle armament was not provided with 360 degree traverse, could not be considered a tank. It is now considered that the restricted traverse on the vehicle should not be the determining factor as to whether the vehicle should be called a tank or a gun motor carriage. Since a gun motor carriage is usualy characterized by the provision of relatively light armor protection and the subject vehicle is provided with a maximum of armor (12" maximum), and therefore satisfies the accepted requirements for a tank, namely, a combination of maximum firepower and maximum armor protection, it is considered advisable to rename this vehicle as a tank."

And so it was done, and to this day, the vehicle is listed as Super-Heavy Tank T28.

Above: These two photographs show the commander's and loader's positions looking back from the driver's and gunner's positions respectively. The photographs are dated October 1946. Below: Showing the gunner's and driver's positions, these undated photographs likely were taken after the conclusion of testing, as evidenced by the removed sighting equipment and the wooden block holding the gun in place.

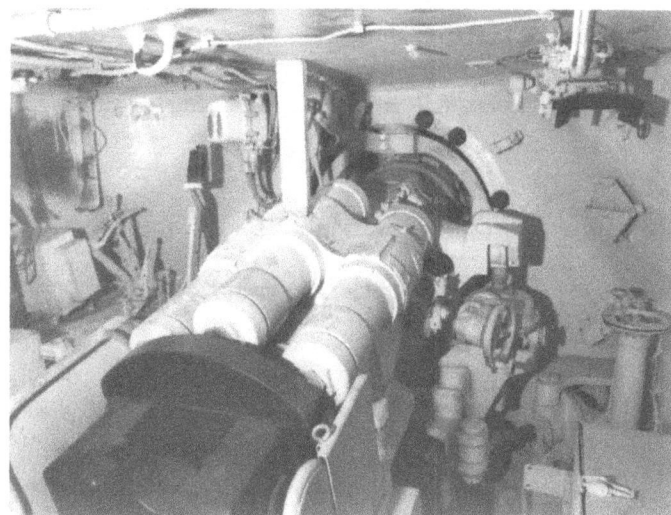

The death knell came in the report from Aberdeen in October of 1947, with the report concluding that with the successful development of the T29 which mounted the same gun in a 360 degree turret, there was no purpose in continuing the tests. However, although the program was cancelled with only two vehicles completed, as the heaviest thing around, the vehicle continued to see service as a load test vehicle. One of these tests was to determine the ability of the US Navy to land 100 ton tanks onto beaches by use of an LST. These tests were conducted in Little Creek, VA, 20-23 April 1948.

The report by Commander, Transport Division Twenty-One which was appended to the Army report, provided an interesting perspective on the Naval requirements, as well as their thought processes.

A point brought up during this conference was: Why conduct a loading test of the Super Heavy Tank, T28, when it has already been declared obsolete? The answer to this question is quickly apparent when the successors to the T28 Tank are viewed. Attached as Enclosure (E) is a photograph of the T29 Tank. Another in production is the T30. No photograph of the T30 Tank was available. The T29 and the T30, however, are identical except for the gun carried.

The T29 mounts a 105 mm high velocity gun. The T30 mounts a high velocity 155mm gun. It will be noted that the T29 weighs approximately 75 tons, just slightly more than the T28 with the outer tracks removed. The experience gained in loading and unloading the T28 would be extremely valuable when we are called upon to transport the T29 and the T30. In gaining this experience, we would be working with an obsolete weapon and, should an accident occur and the tank be damaged, it would be an obsolete weapon which had been damaged and not one of the latest models.

Another point considered during these discussions was: "Why bother with the outer tracks of the T28 when this tank, with the outer tracks removed, weighs almost as much as the T29?" Should the present "Cold War" suddenly develop into a "shooting war" and the need for a weapon such as this arise, a request to transport it to the troubled area would not be based on the fact that it was or was not obsolete. The T28 exists and we should know the problems which exist in transporting it, should an emergency arise.

In view of the foregoing discussion, the following recommendations are submitted:
(a) That the LCTs which are to be used in this test be tested at the earliest practicable date to determine whether or not they can carry such a load, 75 tons, with a ground pressure of twenty (20) pounds per square inch. Tests should include ramp, ramp hinge pins and deck. The test of the deck should be made while the bottom of the LCT is resting on the beach to determine whether or not this weight would crush the double bottoms. These tests should be made on the LCTs now undergoing overhaul at the Norfolk Naval Shipyard.
(b) That a suitable beaching point be developed at Little Creek for conducting this test. This point should include the characteristics enumerated in paragraph 10 of this report. The weight involved is much greater than any previously experienced. The provision of a hard surface on which the tank can move after leaving the ramp should be considered carefully. Existing types of beach roadway should be tested to determine whether or not they will support this weight.
(c) The subject of soil-stabilization should be investigated. This subject is under study in the Bureau of Yards and Docks.

First, T28 with outboard tracks in travel mode, and using a tow bar from an M26 truck, was loaded aboard an FCT(6) at the Aberdeen Proving Ground amphibious ramp, which then embarked upon an LSD waiting off Howell Point. The ship then proceeded to Lynnhaven Roads, where a special ramp had been constructed for the offloading of the tank. After doing so, the tank was then loaded aboard LST-1153, by use of the same ramp. The whole convoluted process was required because the State of Maryland had issues with transporting 100-ton vehicles on their road, and the spring thaws made overland travel difficult (plus it would probably take forever), so it was the quickest and cheapest way of getting the tank to the LST.

This resulted in no great issues, and so the outboard tracks were installed (it took about two hours), and the tank reloaded aboard the LST. Again, the experiment proved successful, although there was a clearance of exactly one inch to each side between the tank and the LSTs guardrails. The next day, the LST proceeded on to a nearby beach, where it was test unloaded again. After it was again reloaded, the LST returned to the pier, and the tank transferred to the LCT(6), outboard tracks in travel mode. The LCT(6) then returned to the LSD, and then back to Aberdeen.

Top left: Loading onto an LCT(6), note the guide wire to assist in keeping the trailer straight. Top right: The vehicle secured aboard. Center left: Unloading from the LCT(6) onto a ramp to transfer to the LST (not the one alongside). Center right: T28 aboard LST-1153. Below left: Unloading from the LST to a floating causeway.

Bottom right: Another use found for T28 was the testing of heavy-load trailers. Here the tank is found mid-loading onto a 10-wheel, 100-ton semi-trailer T79 in April 1948.

Clockwise from top: The trailer mounted on a flatcar, as it was delivered to Aberdeen on 16 January 1946. T95 on test and apparently undergoing some engine work in an undated photograph. T95 view from above, 21 March 1946, and left rear, 23 January 1946. Note how the track hoists are both on the right side in January, but both forward in March.

Vehicle Data Sheets

Finding the exact specifications for these gun motor carriages seems to have proven rather difficult, especially for the earlier ones. Even for standardised equipment such as the 75mm GMC M3, the US Army's products can't seem to necessarily agree, one can often see a conflict between the specifications given in an operator's manual, the 'overview sheets' produced by Ordnance Branch, and the characteristics sheets from testing. This gets even worse when one notes that over the course of a development process, a vehicle can be changed fairly significantly from month to month. Further, "Military Characteristics" in the various project reports are often the design characteristics laid out by Ordnance Committee as goals to reach, not what was actually attained: Sometimes the distinction is not made clear in the writing, however. A particularly egregious example will be visible in the datasheet for 75mm GMC T67: Specifications provided are found in the documentation for Ordnance Branch's history of the GMC's the Palmer Board, which are in places not even close to agreement. Given I have not yet encountered the raw test report from Aberdeen, I have no way of telling which of the two is more accurate for any particular measurement, so I have provided both.

As a result, it is possible that the specifications which follow are in contradiction to other sources (Most particularly, Hunnicutt, but others as well). It may simply have been that those other sources were looking at a different US Army record than I was.

Thus, I've established some ground rules:

When a vehicle was tested by Ordnance Branch, several days would be taken upon receipt of the vehicle to photograph it and take measurements. On occasion, though, these measurements could be taken at later points in the testing process. These measurements would be incorporated into multiple-page characteristics sheets, often times with a specific date marked on them. As a result, any information found on such a characteristics sheet is given primacy over all other sources. If one of the datasheets on the following pages has a date assigned to it after the vehicle name, then that is the specification of that particular vehicle on the date or month that the characteristics sheet was filled out. It may have been different the day before or the day afterwards.

In the absence of a characteristics sheet, the next tier of information is the text of the test report. After that, the one-page characteristics sheets (Form 50 SPOMD), then the manual, then the project overview characteristics. Finally, if I could find nothing in the Ordnance Branch records at all, I simply left the characteristic out, rather than taking a guess.

Sometimes you will find inconsistencies in the formatting. For example, some sheets give 'maximum grade' in percentage, and others in degrees. (100% = 45 degrees). For better or worse, they are given as seen in the documentation.

37mm Gun Motor Carriage T2

Crew   3

Weight (Loaded, without crew)	2910lb
Length	10' 8"
Width	5'
Height	5'3"
Wheelbase	6' 7½"
Tyres	5.50 or 6.00 x16, 4ply. 32psi front and rear

Armament:

Main	37mm gun M3
Elevation	-10° / +15°
Traverse	+/- 30°
Ammunition	40 Rounds

37mm Gun Motor Carriage T2E1

Crew   2

Weight (Loaded, without crew)	2,835lb . T2E1 (Mod) 3,270
Length	10' 8½"
Width	4'8"
Height	4' 3/¾"
Wheelbase	6' 7½"
Tyres	5.50 or 6.00 x16, 4ply. 32psi front and rear

Armament:

Main	37mm gun M3
Elevation	-10° / +15°
Traverse	360°
Ammunition	40 Rounds

Vision & Fire Control
Telescope, M6

Engine	
Make	Continental BY 4112
Type	Inline
Cylinders	4
Displacement	112cu. in.
hp	45 @ 3,500rpm
Torque	83 lb ft

37mm Gun Motor Carriage T8

Crew	3

Weight (Gross)	5,430lb
Length	15' 2"
Width	6' 7"
Height	6' 8"
Ground Clearance	12½"
Tread (Centre to Centre)	68"
Wheelbase	92"
Tyres	9.00 x 20
Pressure at 0" Penetration	35psi
4" Penetration	5.1psi

Armament:
Main	37mm gun M3
Elevation	-10° / +15°
Traverse	360°
Ammunition	100 Rounds

Armour
Gunshield	¼"

Performance		Engine	
Maximum speed on level	59mph	Make	Ford
Maximum grade	60%	Type	L inline
Fording Depth	25¼"	Cylinders	6
Turning Diameter	52'	Displacement	226 cu. in.
Fuel Capacity	19 gal	hp	90 @ 3,300rpm
Cruising Range	295 miles	Torque	180lb ft @ 1,200rpm
Angle of Approach	53°		
Angle of Departure	36.5°		

Vision & Fire Control
Telescope, M6

37mm Gun Motor Carriage T21 (24NOV41)

Crew   4

Weight (Loaded, with crew)	6,860lb
Weight (Empty)	5,670lb
Length (Excl gun 2' overhang)	14' 6⅝"
Width	6' 8¾"
Height	6' 9¾"
Ground Clearance	11⅛"
Tread (Centre to Centre)	64¾"
Wheelbase	98"
Tyres	9.00x16, 8-ply
Ground Pressure 0 Pen	45psi
4" Penetration	10.6psi

Armament:
Main	37mm gun M3 on Pedestal Mount APG Drawing 8476
Elevation	-10° / +15°
Traverse	360°
Ammunition	80 Rounds

Armour
Gunshield   ¼"

Performance		Engine	
Maximum speed on level	60mph	Make	Dodge T-214
Maximum grade	50%	Type	Poppet valve inline
Turning Diameter	49'	Cylinders	6
Fuel Capacity	30 gal (72 oct)	Displacement	230.2 cu. in.
Cruising Range	180 miles	hp	101 @ 3,200rpm
Angle of Approach	38°	Torque	185lb ft @ 1,600rpm
Angle of Departure	31°		
Vertical Wall	Failed 12"		

Vision & Fire Control	Communications
Telescope M6	Radio Set: SCR-510

37mm Gun Motor Carriage M6 (30 July 1942)

Crew	4

Weight (Gross)	7,350lb
Length	14' 9⅛"
Width	6' 10¾"
Height	5' 1½"
Ground Clearance	10⅝"
Tread (Centre to Centre)	64¾"
Wheelbase	98" 9.00x16,
Tyres	9.00 x16 8-ply

Armament:

Main	37mm gun M3 on Pedestal Mount M25 or M26
Elevation	-10° / +15°
Traverse	360°
Ammunition	80 Rounds
Secondary	Provision for 1x Cal. .30 M1903A1, 3x Cal .30 carbines 12 hand grenades, 10 rifle grenades

Armour	
Gunshield	¼"

Performance		Engine	
Maximum speed on level	55mph	Make	Dodge T-214
Maximum grade	60%	Type	L inline
Speed on 10% grade	20	Cylinders	6
Fording Depth	32"	Displacement	230.2 cu. in.
Turning Diameter	43' 4"	hp	92 @ 3,300rpm
Fuel Capacity	30 gal (72 oct)	Torque	180lb ft @ 1,200rpm
Cruising Range	255 miles		
Angle of Approach	36.5°		
Angle of Departure	31°		

Vision & Fire Control	Communications
Telescope M6	Radio Set: SCR-510

37mm Gun Motor Carriage T14 (15 April 1942)

Crew	4
Weight (Gross)	4,950lb
Weight (Empty)	2,845lb
Length	13' 11⅜"
Width	5'
Height	40⅝"
Ground Clearance	8¾"
Tread (Centre to Centre)	48"
Wheelbase	82" - 36"
Tyres	6.00x16 (Firestone)
Ground Pressure 4" penetration	5 lb /sq in.
Armament:	
Main	37mm gun M3
Elevation	-10° / +15°
Traverse	360°
Ammunition	66 Rounds

Armour	
Gunshield	¼"

Performance		Engine	
Maximum speed on level	55mph	Make	Willys Overland
Maximum grade	50%	Type	L-Head inline
Maximum Speed 20% slope	10mph	Cylinders	4
Fording Depth	24"	Displacement	134.2 cu. in.
Turning Diameter	45' 9"	hp	60 @ 3,600rpm
Fuel Capacity	20 gal (70 oct)	Torque	105lb ft @ 2,000rpm
Cruising Range	275 miles		
Angle of Approach	45°		
Angle of Departure	37°		

37mm Gun Motor Carriage T33

Crew   4

Weight (Loaded)	5,842lb
Weight (Empty)	4,742lb
Length	12' 6¼""
Width	6' 4"
Height	6' 8"
Ground Clearance	10⅞"
Wheelbase	92"
Tyres	9.00 x 16, 6-ply

Armament:
Main	37mm gun M3
Elevation	-10° / +15°
Traverse	360°
Ammunition	80 Rounds

Armour
Gunshield   ¼"

Performance		Engine	
Maximum speed on level	59mph	Make	Ford
Turning Diameter	52'	Type	L inline
Fuel Capacity	34 gal	Cylinders	6
Cruising Range	360miles	Displacement	226 cu. in.
Angle of Approach	40°	hp	90 @ 3,300rpm
Angle of Departure	35°	Torque	180lb ft @ 1,200rpm

37mm Armored Car T23

Crew 4

Weight (Gross)	13,500lb
Length	16'8"
Width	7'6"
Height	6' 7¾"
Wheelbase	118"
Tyres	12.00x20

Armament:

Main	37mm Gun M5
Traverse	360

Secondary	2x Cal. .50 M2HB

Armour		Turret	
Upper Front Hull	½"	Front	?"
Sides	⅜"	Side	½"
Rear	⅜"	Rear	½"

Engine

Make	
Type	Inline Gasoline
Cylinders	6
Displacement	331cu. in.
hp	104 @ 2,800rpm
Torque	237 lbs ft @ 800rpm

Scout Car M3A1E3 with 37mm Gun (9 July 1941)

Crew 4

Weight (Gross)	??lb
Length	18'6"
Width	6' 5¼"
Height	8' ¾"
Ground Clearance	8 ¾"
Tread (Centre to Centre)	63¼"
Wheelbase	131¼"
Tyres	8.25x20, 10-ply
Ground Pressure. 0" Pen	60 psi
4" Pen	20.4psi

Armament:

Main	37mm Gun M3 on Pedestal Mount T-6
Elevation	+15° / -19°
Traverse	Left 90°, right 43°
Ammunition	40
Secondary	2x Cal. .30 M1917, 8,000 rounds

Armour

Upper Front Hull	¼"
Sides	⅜"
Rear	⅜"
Top	Open

Performance		Engine	
Maximum Speed on level	50mph	Make	White 160
Maximum Grade	60%	Type	L-Head Inline
Trench Crossing Ability	1'6"	Cylinders	6
Vertical Obstacle	12"	Displacement	320 cu. in.
Fording Depth	28"	hp	87 @ 2,400
Turning Diameter	57'	Torque	220 lb ft @ 1,150
Fuel Capacity	54USG		
Cruising Range	250miles		

57mm Gun Motor Carriage T44

Crew   3

Weight (Gross)	7,150lb
Weight (Empty)	5,850lb
Length	12' 6¼"
Width	6' 4"
Height	4'9'"
Ground Clearance	10.87"
Wheelbase	92"
Ground Contact Length	83"
Tyres	9.00x16, 6-ply

Armament:

Main	57mm Gun M1
Elevation	-10° / +15°
Traverse	45° left / 45° right
Ammunition	56 Rounds

Gunshield Armour

Front	⅜"
Sides	¼"

Performance		Engine	
Maximum Speed on level	59mph	Make	Ford
Fuel Capacity	34 gal	Type	
		Cylinders	6
		Displacement	226 cu. in.
		hp	90 @ 3,300rpm
		Torque	180 lb ft @ 1,200rpm

75mm Gun Motor Carriage T27 (25 October 1941)

Crew   Undetermined

Weight (Empty)	9,350lbs
Length	15' 8"
Width	7'8"
Height	7' 6" (4.2" without shield)
Ground Clearance	16"
Wheelbase	10' 4"
Tyres	11.00 - 20 Single

Armament:

Main	75mm Gun M1897A4
Elevation	-8°/ +25.5°
Traverse	60°
Ammunition	Undetermined

Performance

Maximum speed on level	55mph
Fuel Capacity	36.5 gal
Angle of Approach	45°
Angle of Departure	42°

Other characteristics same as for 1½ ton, 4x4, Studebaker Truck.*

*The document in the archives contains a handwritten comment that there is more than one Studebaker 4x4 1 ½ ton truck, but no response indicates which one (eg M15A or M16) was appropriate

75mm Gun Motor Carriage T66

Crew   5

Weight (Gross)	31,500lb
Weight (Empty)	28,000lb
Length (Over gun/Hull)	19' 6" / 18' 4"
Width	10'
Height	7' 5"
Ground Clearance	16"
Tread (Centre to Centre)	102"
Wheelbase	126" (63"+63")
Tyres	14.00x20
Ground Pressure at 3"	12psi

Armament:

Main	75mm Gun M3 on Mount M34 (Modified)
Elevation	-10° / +20° *
Traverse	360°
Ammunition	63 Rounds

Secondary	Cal. .50 M2HB Flex, 300 rounds.
	Provision for 5x .30 Carbine, 500 rounds. 16 Grenades

Armour

Upper Front Hull	½" @ 45°		
Sides	⅜" @15°	Turret	65" Ring
Rear	⅜" @60° and 0°	Front	½" @ 22°
Top	¼"	Side	⅜" @ 22°
Bottom	¼"	Rear	⅜" @ 22°

Performance		Engine	
Maximum speed on level	57mph	Make	Cadillac
Maximum grade	60%	Type	Twin Series 42 V
Speed on 3% grade		Cylinders	2x 8
Trench crossing ability	5'3"	Displacement	792cu. in.
Vertical Obstacle	24"	hp	220hp @ 3,700
Fording Depth	32"	Torque	488 lb ft @ 1,200rpm
Turning Diameter	66'		
Fuel Capacity	106 gal (80 Oct)		
Cruising Range	300 miles		

Vision & Fire Control	Communications	* Ordnance Branch project report (and Hunnicutt)
Periscopes, M6	Radio Set: SCR-506	states +45° , but the Ordnance document covering
Telescope M54	and	75mm GMC history and related T67 GMC
Telescope M40	SCR 608 or SCR 610	information indicates +20° is the correct figure

3" Gun Motor Carriage T7 (Proposed Specifications)

Crew   5

Weight (Gross)	11-12 tons
Weight (Empty)	
Length / Over gun	17'
Width	8' 4"
Height (min practicable)	7'9"

Armament:

Main	3" Gun M6
Elevation	-10° / +15°
Traverse	15° left / 15° right
Ammunition	50 rounds

Secondary

Armour

Front	1"
Side Wings	½"
Top	½"

Performance

Maximum Speed	45mph
Speed on 10% grade	15mph

Vision & Fire Control
Telescope, M6

3-inch Gun Motor Carriage T55

Crew   5

Weight (Gross)	38,300lb
Length (Over gun/Hull)	25' 7½" / 17' 11"
Width	9' 3⅝"
Height	8' 7½"
Ground Clearance	15"
Tread (Centre to Centre)	93"
Tyres	14.00 x 20

Armament:

Main	3" Gun M6 in Mount M4
Elevation	+15° / -10°
Traverse	20° left, 20° right
Ammunition	55 Rounds
Secondary	Pedestal mount in rear compartment

Armour

Upper Front Hull	¼"
Sides	¼"
Rear (Upper)	¼"
Rear (Lower)	⅛"
Top	⅛"
Bottom	⅛"

Engine

Make	Cadillac
Type	Twin Series 42 V
Cylinders	2x 8
Displacement	792cu. in.
hp	220hp @ 3,400rpm
Torque	448lb ft @ 1,200rpm

Vision & Fire Control

Communications
Radio Set: SCR 610

3-inch Gun Motor Carriage T55E1

Crew   5

Weight (Gross)	30,200lb
Length (Over gun/Hull)	24' 7" / 17' 11"
Width	9' 3½'
Height	6' 3⅝"
Ground Clearance	15"
Tread (Centre to Centre)	93"
Wheelbase	170"
Tyres	14.00 x 20
Ground Pressure 4"	9.5 psi

Armament:

Main	3" Gun M7 in Mount M4
Elevation	+17° (Centerline) / -10°
Traverse	20° left, 20° right
Ammunition	45 Rounds
Secondary	Pedestal mount in forward compartment

Armour

Upper Front Hull	¼"
Sides	¼"
Rear (Upper)	¼"
Rear (Lower)	⅛"
Top	⅛"
Bottom	⅛"

Performance		Engine	
Maximum Speed on level	50mph	Make	Cadillac
Maximum Grade	62%	Type	Twin Series 42 V
Speed on 3% grade		Cylinders	2x 8
Trench Crossing Ability	38"	Displacement	792cu. in.
Vertical Obstacle	40"	hp	220hp @ 3,400rpm
Fording Depth	36"	Torque	448lb ft @ 1,200rpm
Turning Diameter	98' 7½"		
Fuel Capacity	50usg		
Cruising Range	150 miles		

Communications
Radio Set: SCR 610

75mm Gun Motor Carriage M3 (M3A1)

Crew   5

Weight (Gross)	20,000lb
Length	20' 5½"
Width	7'1"
Height	8' 2⅝"
Ground Clearance	11 3/16"
Tread (Front)	64½"
Tread (Rear)	63 13/16"
Wheelbase	135½"
Ground Contact Length (Tracks)	46¾"
Tyres	8.25x20, 12-ply

Armament:

Main	75mm Gun M1897A4 on Mount M3 or (Mount M5)
Elevation	-10° / +29° (-6.5° / +29°)
Traverse	19° left / 21° right (21° left, 21° right)
Ammunition	59 Rounds

Secondary: Provision for 1 cal. .30 Rifle, M1903, 4x carbine cal. .30
12 hand grenades (5 Frag, 5 smoke, 2 incend.), 10 rifle grenade

Armour		Gunshield Armour	
Front	¼"	Front	⅝"
Sides	¼"	Sides	¼"
Rear	¼"	Top	¼"
Windshield	½"		

Performance		Engine	Yes
Maximum speed on level	45mph	Make	White 160
Maximum Grade	60%	Type	L-Head Inline
Speed on 4% Grade	25 mph	Cylinders	6
Vertical Obstacle	12"	Displacement	386 cu. in.
Turning Diameter	60'	hp	147 @ 3,000
		Torque	325 lb ft @ 1,200

Vision & Fire Control	Communications
Telescope M33	Radio Set: SCR 510

57mm Gun Motor Carriage T48 (August 1942)

Crew   5

Weight (Gross)	19,000lb
Length	21' ⅝"
Width	7'1"
Height	7'
Ground Clearance	11' 3/16"
Wheelbase	135½"
Ground Contact Length (Tracks)	46¾"
Tyres	8.25x20, 12-ply

Armament:

Main	57mm Gun M1 on Mount T5
Elevation	-7.5° / +15°
Traverse	27.5° left / 27.5° right
Ammunition	100 Rounds

Secondary: Provision for 5 cal. 303 Rifle, British
                    22 rifle grenades

Armour		Gunshield Armour	
Front	¼"	Front	⅝"
Sides	¼"	Sides	¼"
Rear	¼"	Top	¼"
Windshield	½"		

Performance		Engine	
Maximum Speed on level	45mph	Make	White 160
Maximum Grade	75%	Type	L-Head Inline
Speed on 4% Grade	25 mph	Cylinders	6
Angle of Approach	37 deg	Displacement	386 cu. in.
Angle of Departure	32 deg	hp	147 @ 3,000
Vertical Obstacle	12"	Torque	325lb ft @ 1,200
Fording Depth	32"		
Turning Diameter	60'		
Fuel Capacity	60 gal		
Cruising Range	200 miles		
Approach Angle	37°		
Departure Angle	32		

Vision & Fire Control	Communications
Telescope, M18	British Wireless #19

57mm Gun Motor Carriage T49

Crew   5

Weight (Gross)	32,000lb
Length	15' 2"
Width	7' 8"
Height	7' 3"
Ground Clearance	14"
Ground Pressure	10 psi
Ground Contact Length (Tracks)	9' 8¼"
Tread	89¼"

Armament:

Main	57mm Gun M1
Elevation	-10° / +45°
Ammunition	49 Rounds

Secondary:
2x Cal. .30 M1919A4 (Coax & Bow), 2,500 rounds
1x Cal. .45 M1928A1 SMG, 2,000 rounds
1x Cal. .30 M1903 rifle. 24x hand grenades

Armour		Turret Armour	
Front	¾"	Front	¾"
Sides	⅜"	Sides	¾"
Bottom	⅜"	Rear	¾"
Top	⅜"		

Performance

Maximum Speed on level	55mph
Maximum Grade	60%

		Engine Make	Buick 60 Series
Vertical obstacle	12"	Type	Overhead Valve
Fording Depth	27"	Cylinders	8
		Displacement	
Fuel Capacity	115 gal	hp	144 @ 3,700rpm
Cruising Range	230 miles	Torque	243lb ft @ 2,500
Trench Crossing	7'		
		Transmission	Torque Converter
			3 Fwd/1 Reverse

Communications
SCR-508

75mm Gun Motor Carriage T67
Data from Special Armored Vehicles Board chart, (Data from Ordnance History File, if different)

Crew   5

Weight (Gross)	32,000lb
Length	17' 10½" (17' 4½")
Width	8' 9¾" (8' 9½")
Height	7' ¼"
Ground Clearance	15" (14½")
Ground Pressure	11.34 psi
Ground Contact Length (Tracks)	9' 6"
Track width	12"
Tread	(89¼")

Armament:

Main	75mm Gun M3 in Combination Mount M34 (Modified)
Elevation	-5° 20' / +20° (-10° / +20°)
Traverse	Power
Ammunition	41 (40) Rounds

Secondary:	(Provision for Cal. .50 M2HB)
	(Provision for 1x Cal. .30 M1903, 4x Cal. 45 SMGs)
	300 cal. 45 for SMGs

Armour		Turret Armour	
Front	½"	Front	½" (1" With Gunshield)
Sides	½"	Sides	½"
Rear	½"	Top	Open
Top	5/16"		

Performance

Maximum Speed on level	51 (50) mph		
Maximum Speed on 3%	(30mph)		
Maximum grade	(60%)	Engine Make	Buick
Vertical Obstacle	24" (12")	Type	Inline Watercooled
Fording Depth	32"	Cylinders	8
		Displacement	386 cu. in. (320 cu. in.)
Fuel Capacity	118 gal	hp	330 @ 3,800rpm
Cruising Range	250 miles (150miles)	Torque	382lb ft @ 2,600
Trench Crossing	6' 4"		
		Transmission	Torque-Matic
			3 Fwd/1 Reverse

Vision & Fire Control
Telescope, M18
(Telescope, M54, Periscope M4
w/ Telescope M40)

Communications:
(SCR 508 or 510)

75mm Gun Motor Carriage M8 w/ 75mm M3 Tank Gun

Crew   4 or 5

Weight (Empty)	30,205lb
Length	14' 6¾"
Width	7' 4½"
Height	7' 6½"
Ground Clearance	8' 1¼"
Ground Pressure	11.83psi

Armament:
Main	75mm Gun M3
Elevation	
Traverse	
Ammunition	44 Rounds

Armour		Turret Armour	
Front	1½" max	Front	1½"
Sides	1⅛"	Sides	1"
Rear	¼"		

Performance		Engine	
Maximum Speed on level	46mph	Make	2x Cadillac Series 42
Maximum Grade	30°	Type	Liquid-cooled
Gap Crossing	5' 6"	Cylinders	2x V8
Vertical Obstacle	24"	Displacement	2x 346 cu. in.
Fording Depth	36"	hp (net)	2x 110 @ 3,400rpm
Turning Radius	13'	hp (gross)	2x 148hp @ 3,200rpm
Fuel Capacity	178 gal, 80 Oct.	Torque (Net)	2x 244lb ft @ 1,200
Cruising Range	175 miles		

3-inch Gun Motor Carriage M5 (01 September 1942)

Crew   4

Weight (Loaded)	22,570lb
Length	15' 1""
Width	8' 3"
Height	6' 2'"
Ground Clearance	16"
Track Width	14"
Ground Pressure	9.2psi
Tread	78"

Armament:

Main	3" Gun M6
Elevation	-10°/+15°
Traverse	11° left /18° right **
Ammunition	33 Rounds

Secondary:

Armour

Max	½"
Min	⅜"

Performance		Engine	
Maximum Speed on level	38 mph*	Make	Hercules
Maximum Speed in reverse	4.3 mph	Type	Watercooled Diesel
Speed on 4% grade	25 mph	Cylinders	6
Angle of Approach	40 deg	Displacement	404 cu. in.
Angle of Departure	45 deg	hp	150 @ 2,800
Vertical Obstacle	12"	Torque	316lb ft @ 1,950
Fording Depth	30"		
Turning Diameter	34'		
Fuel Capacity	62 gal		
Cruising Range	192 miles		

Vision & Fire Control	Communications
Telescope, M41	SCR 510

*A later test run reached 42mph
**Removing the gunner's seat increased this to 23°

3-inch Gun Motor Carriage T24

Crew   6

Weight (Gross)	54,600lb
Length	18' 7¾"
Width	8' 11⅜"
Height	8' 7"
Ground Clearance	19"
Ground Pressure	11.6 psi
Ground Contact Length (Tracks)	12' 3"
Tread	83"

Armament:

Main	3-Inch AA Gun M3 in Mount M2A2
Elevation	-2° / +15° (But see text)
Traverse	33°
Ammunition	40 Rounds

Armour

Front	2"
Sides	1"
Rear	¾"

Performance		Engine	As per M3 Medium
Maximum Speed on level	24mph		
Maximum Cross-country	15mph		
Maximum Crade	60%		
Vertical Obstacle	24"		
Fording Depth	32"		
Turning Diameter	68'		

Vision & Fire Control
Telescope, M18

3-Inch Gun Motor Carriage T56

Crew   5

Weight (Gross)	32,000lb
Length	16' 10¾"
Width	8' 3"
Height	7' 9½"
Ground Clearance	14"
Tread	73¼"
Ground Pressure	19.6 psi
Ground Contact Length (Tracks)	121½"

Armament:

Main	3" Gun M7 on pedestal mount
Elevation	-5° / +25°
Traverse	15° right and 15° left
Ammunition	40 Rounds

Secondary:

1x Cal. .30 MG M1919A4, 500 rounds
Provision for 2x Cal. .30 Carbines M1, 100 rounds
8x M8 (Smoke) Hand Grenades, 4x Drop-type smoke pots

Armour		Gun Shield	
Front	1½"	Front & Sides	1½"
Sides	1⅛"	Top	½"
Rear	½"		
Top	⅜"		
Bottom	⅜"-½"		

Performance		Engine	
Maximum speed on level	40mph	Model	Continental W670-Series 12
		Type	Radial
		Cylinders	7
Fuel Capacity	50 gal., 80 Octane	Gross hp	288 @ 2,600rpm
Cruising Range	70 miles	Torque	605 lbft @ 2,200 rpm

Communications: SCR 510		Transmission	Synchromesh

3-Inch Gun Motor Carriage T57

Crew   5

Weight (Gross)	33,000lb
Length	16' 10¾"
Width	8' 3"
Height	7' 9½"
Ground Clearance	14"
Tread	73¼"
Ground Pressure	19.6 psi
Ground Contact Length (Tracks)	121½"

Armament:
Main	3" Gun M7 on pedestal mount
Elevation	-5° / +25°
Traverse	15° right and 15° left
Ammunition	40 Rounds

Secondary:
1x Cal. .30 MG M1919A4, 500 rounds
Provision for 2x Cal. .30 Carbines M1, 100 rounds
8x M8 (Smoke) Hand Grenades, 4x Drop-type smoke pots

Armour
Front	1½"
Sides	1⅛"
Rear	½"
Top	⅜"
Bottom	⅜"-½"

Performance		Engine	
Maximum Speed on level	40mph	Model	Wright R975-C1
Fuel Capacity	50 gal., 80 Octane	Type	Radial
Cruising Range	70 miles	Cylinders	9
		Gross hp	400 @ 2,400rpm
		Torque	900 lbft @ 2,000 rpm

Communications: SCR-510		Transmission	Synchromesh

3-Inch Gun Motor Carriage M10 (From TM 9-752)

Crew   5

Weight (Combat)	66,000lb
Length	20' 2"
Width	10'
Height	8' 1½"
Ground Clearance	18"
Tread	83"
Ground Pressure	
Ground Contact Area	3,346 sq. in.

Armament:

Main	3" Gun M7 in Mount M5. Telescope M51
Elevation	-10° / +30°
Traverse	Manual
Ammunition	54 Rounds

Secondary:

1x Cal. .50 MG M2, 350 rounds
5x Cal. .30 Carbines M1, 300 rounds
1x Submachine Gun, Cal. .45, M1928A1, 510 rounds

Armour			Turret	
Front	1½" @ 54°		Front	2.25" @ 45°
Upper Sides	¾" @ 38°		Sides	1"
Lower sides	1"		Top	3/4"
Top	⅜"			
Rear	¾" @ 30°			
Skirts	¼" @ 35°			
Appliqué	¼" - 1"			

Performance

Maximum Speed on level	30mph		Engine	
Reverse	3mph		Model	GM Series 71, Model 6046
Trench Crossing	7' 5"		Type	Twin inline diesel
Vertical Obstacle	18"		Cylinders	2x6
With grousers	36"		Displacement	2x425 cu. in.
Maximum Fording Depth	40"		Rated hp	375hp @ 2,100rpm
Turning Radius, 1st Gear	26'		Torque at Shaft	800 lbft @ 1,800rpm
Turning Radius, 5th Gear	50'			
Maximum Grade	27°		Communications: SCR 610, Telephone RC99	
Fuel Capacity	167* gal. 50 Cetane			
Cruising Range	140 miles (18mph)		Transmission	Buick Synchromesh 5-speed

* From Tabulated Data page. Bizarrely, the TM later on lists a total of 148 gal. Test reports indicate 164-167.

76mm Gun Motor Carriage T70 (Armored Force Board 27 December 1943)

Crew 4

Weight (Combat)	37,900lb
Length	17' 4"
Width	8' 5"
Height	8' 5 7/16" (Over AA Gun)
Ground Clearance	14¼"
Tread	94.6"
Ground Pressure	12.5psi (Zero penetration), 11.8psi (3" Penetration)
Track Width	12", 5" Pitch

Armament:

Main	76 mm Gun M1
Traverse	Power
Ammunition	45 Rounds

Secondary:
1x Cal. .50 MG M2, 1000 rounds
4x Cal. .30 Carbines M1, 460 rounds

Armour		Turret	
Front	½"	Front	½"
Floor	¼"		

Performance

Maximum speed on level	50mph
Maximum Grade	50%
Fuel Capacity	160gal. 80 Octane

Engine*

Model	Continental R 975-C1
Type	Radial
Cylinders	9
Displacement	973 cu. in.
Rated hp	400 @ 2,400rpm
Torque at Shaft	900 lbft @ 1,800rpm

Communications: SCR 610, Telephone RC99

Transmission	Torqmatic (3-Speed)

*Note that hp and torque will differ from R 975-C1 on M18 page. The figures are suspect, but are as reported

76mm Gun Motor Carriage M18 (TM 9-755)

Crew   5

Weight, Fighting (Less Crew)	36,510lb
Length	21' 10"
Width	9' 5"'
Height	8' 5" (w/ AAMG), 7' 9¼" Minimum
Ground Clearance	14¼"
Tread	7' 10⅝"
Ground Pressure	11.9 psi
Ground Contact Area	3,156 sq. in.

Armament:

Main	76mm Gun M1A1, M1A1C or M1A2
Elevation	-10° / +20°
Traverse	Hydraulic
Ammunition	45 Rounds

Secondary:
1x Cal. .50 MG M2, 800 rounds
5x Cal. .30 Carbines M1, 450 rounds
6x Grenades, Smoke, WP M50, 6x Grenades, Frag, MkII. 4x Smoke Pot

Armour		Turret	
Front (Upper)	½" @ 64° & 38°	Front (Shield)	¾"
Front (Lower)	½" @ 24° & 53°	Sides	½"
Upper Sides	½" @ 23°	Rear	½"
Lower sides	½"		

Performance		Engine	
Maximum Speed on level	45mph	Model	Continental R 975-C1 or C2*
Reverse	20mph	Type	Radial
Maximum Fording Depth	48"	Cylinders	9
Turning Radius	33'	Displacement	973 cu. in.
Maximum Grade	60% @ 3mph	Rated hp	350 (C1) or 400 (C2) @2,400rpm
Fuel Capacity	165 gal. 80 Oct	Torque at Shaft	800 lbft @ 1,800rpm
Angle of Approach	28°		
Angle of Departure	26½°		

Communications: SCR 610, Telephone RC99

Transmission	Torqmatic (3-speed)

76mm Gun Motor Carriage T72

Crew   5

Weight (Combat)	
Length	20' 2"
Width	10'
Height	
Ground Clearance	18"
Tread	83"
Ground Pressure	
Ground Contact Area	3,346 sq. in.

Armament:	
Main	76mm Gun M1 in Mount T2
Elevation	-10° / +30°
Traverse	Manual
Ammunition	97 Rounds

Secondary:	
	1x Cal. .50 MG M2, 350 rounds
	5x Cal. .30 Carbines M1, 300 rounds
	1x Submachine Gun, Cal. .45, M1928A1, 510 rounds

Armour		Turret	
Front	1½" @ 54°	Front	1.5"
Upper Sides	¾" @ 38°	Sides	1⅛"
Lower sides	1"	Rear	1⅛"
Top	⅜"	Top	Open
Rear	¾" @ 30°		
Skirts	¼" @ 35°		

Performance			
Maximum speed on level	30mph		
Reverse	3mph	Engine	
Trench Crossing	7' 5"	Model	Ford GAA
Vertical Obstacle	18"	Type	Water-cooled V8
With grousers	36"	Cylinders	8
Maximum Fording Depth	40"	Displacement	1,100 cu. in.
Turning Radius, 1st Gear	26'	Rated hp	450 @ 2,600 rpm (Net)
Turning Radius, 5th Gear	50'	Torque	950 lbft @ 2,200rpm (Net)
Maximum Grade	27°		
Fuel Capacity	192gal. 80 Octane		
Cruising Range	150 miles	Transmission	Synchromesh 5-speed

76mm Amphibious Motor Carriage T86 / (T86E1)

Crew   5

Weight (Combat)
Length	30' 7" (27'4")
Width	10' 2"
Height	9' 6"
Ground Clearance	16" ¼"
Tread	94⅝"
Track Width	21"
Ground Contact Length	116¼"

Armament:
Main	76mm Gun M1
Elevation	
Traverse	Manual
Ammunition	100 Rounds
Secondary:	1x Cal. .50 MG M2, 800 rounds

Armour   1/4"

Performance		Engine	
Maximum speed on level	43mph	Model	Continental R 975*
Max speed on water	5.1mph (6.2mph)	Type	Radial
Maximum Grade	60%	Cylinders	9
Fuel Capacity	170gal. (225 gal)	Displacement	973 cu. in.

Communications: SCR 610, British #19 **

* Documentation is non-specific as to the exact variant of R975 used.
** No explanation given for the British radio, but that's what the sheet says.

90mm SP Gun T101 (29 MAR 55)

Crew   4

Weight (Combat, with crew)	15,900
Length	20' 1 11/16"
Width	8'2" (Reducible to 8')
Height	86½"
Ground Clearance	12½"

Armament:
Main	76mm Gun T125
Elevation	-10° / +15°
Traverse	Manual, 30° left & Right
Ammunition	29 Rounds

Fuel Capacity   32 gal.

Communications: AN/PRC 8, 9, 10

Vision/Fire Control: Telescope, T152

90mm SP Gun M56 (21 July 1958 DA Form 4288)

Crew   4

Weight, Combat (Empty)	15,750lbs (12,500lbs)
Length	19' 2"
Width	8' 5½"
Height	6' 9"
Ground Clearance	12¾"
Tread	78"
Track Width	20"
Ground Pressure	4.25psi
Tyres	7.5x12

Armament:	
Main	90mm Gun M54 in Mount M88
Elevation	-10° / +15°
Traverse	30° Left/ 30° Right
Ammunition	29 rounds

Secondary:	Provision for 4x Carbine, Cal. .30, 240 rounds, 8x Grenades

Armour: None (Of course, the gunshield is metal, but the datasheet apparently doesn't consider it armour)

Performance

Maximum speed on level	28mph		
Vertical Wall	30"	Engine	
Trench Crossing	48"	Model	Continental AOI-402-5
Maximum Fording Depth	42"	Type	Air-cooled, opposed
With Kit.	60"	Cylinders	6
Turning Radius	6½'	Displacement	402 cu. in.
Maximum Grade	60%	Rated hp	200 @ 3,000 rpm gross (165 net)
Fuel Capacity	50gal. 83 Octane	Torque	347lbft @ 2,800rpm gross (325 net)
Cruising Range	140 miles		

Communications: AN/PRC-10   Transmission   Cross-Drive Torque Converter CD-150-4

Fire Control: Telescope, T186 in Mount T129

105mm Gun Motor Carriage T95 (T28 Pilot #1, 12 AUG 1947)

Crew   4

Weight (Combat)	190,000lbs
Length	25' 2½"
Width	14' 11" (Transport: 10'4")
Height	8' 10" (Over Cupola), 9'4" (Over MG)
Ground Clearance	19"
Tread	126" (104½" for transportation)
Ground Pressure	11.7psi @ zero penetration
Ground Contact length	17' 6"

Armament:

Main	105mm Gun T5E1 with Sight, Periscope, M10E3, Sight Telescope T1939
Elevation	-4° / +18° 42'
Traverse	11° Left/ 13° Right
Ammunition	62 rounds

Secondary:	1x Cal. .50 MG M2, 660 rounds

Armour		Mantlet	
Front	12" @ 0°	Front	11.5-12"
Lower Front Hull	5¼" @ 58°		
Upper Hull Sides	2½" @ 58°		
Lower Hull Sides	1½"		
Rear	2" @ 9°		
Top	1½"		
Bottom	1"		
Skirting	4"		

Performance		Engine	
Maximum speed on level	8mph		
Cross Country	4mph	Engine	
Trench Crossing	8' 2½"	Model	Ford GAA
		Type	Water-cooled V8
Maximum Fording Depth	48"	Cylinders	8
Turning Diameter	70'	Displacement	1,100 cu. in.
Turning Radius, 5th Gear		Rated hp	500 @ 2,600 rpm
Maximum Grade	60%	Torque	1050 lbft @ 2,200rpm
Fuel Capacity	400gal. 80 Octane		
Cruising Range	100 miles		
		Transmission	Torqmatic 3-speed

www.ingramcontent.com/pod-product-compliance
Lightning Source LLC
Chambersburg PA
CBHW040141200326

41458CB00025B/6333